The Arctic Council

Svein Vigeland Rottem

The Arctic Council

Between Environmental Protection and Geopolitics

palgrave
macmillan

Svein Vigeland Rottem
Fridtjof Nansen Institute
Lysaker, Norway

ISBN 978-981-13-9292-4 ISBN 978-981-13-9290-0 (eBook)
https://doi.org/10.1007/978-981-13-9290-0

This Palgrave Pivot imprint is published by the registered company Springer Nature Singapore Pte Ltd.
The registered company address is: 152 Beach Road, #21-01/04 Gateway East, Singapore 189721, Singapore

Contents

CHAPTER 1

The Arctic Council: From Environmental Protection to Geopolitics

Abstract This chapter introduces the topic and the structure of the book and outlines to the history of the Arctic Council, starting from a limited effort to promote cooperation on environmental policy to becoming the most important international forum for discussions on a wide range of Arctic issues. It also portrays the political and geographical landscape the Arctic Council operates in.

Keywords Arctic Council • Climate change • Conflict • Cooperation

In consequence of a changing climate, the Arctic has once again secured a place on the international agenda. It is in the Arctic global climate change is most conspicuous. Higher temperatures and diminishing ice cover are affecting the basic conditions for actors with interests in the region. Technological progress has also made it possible to extract resources in areas that were previously inaccessible.

Commercial activity is increasing and each of the Arctic states wants to take part in shaping the "new" Arctic. In parallel, a number of non-Arctic states also have aspirations for the region. The agenda varies from research to geopolitics, but that something is happening in the Arctic that needs careful monitoring is becoming clear to increasing numbers.

Although interest in the region is growing, it is still characterised by political stability and well-functioning governance regimes. Among the

S. V. Rottem, *The Arctic Council*,
https://doi.org/10.1007/978-981-13-9290-0_1

Arctic states, there is consensus on the basic rules. That is not to deny the existence of different approaches to what a future Arctic should look like, from ideas of total protection to expansive oil and gas extraction. The Arctic's position on the political agenda also varies among the Arctic coastal states, from Norway's High North hype to a more peripheral place in the foreign policy of the US.

It is in this political and geographical landscape the Arctic Council operates. All the Arctic states describe the Council as the most important international forum for dialogue and cooperation on Arctic challenges and opportunities, and as a significant arena for developing knowledge and learning about a changing region. Despite portrayals like these in addresses to the public and official political documents, knowledge of the forum is actually in short supply. In a survey conducted in 2015, in which residents from all the Arctic states participated, fewer than half were aware of the Arctic Council.[1] The purpose of this book is therefore to introduce the reader to the Arctic Council and its role as a platform for dealing with local, national, regional and global challenges of relevance to the "new" Arctic.[2]

The book is divided into five chapters. Chapter 1 presents the history of the Arctic Council, starting from a limited effort to promote cooperation on environmental policy to becoming the most important international forum for discussions on a wide range of Arctic issues. Chapter 2 introduces the Council's organisation and gives a short introduction to the Arctic profile of the member states. This is necessary because the states decide the parameters for cooperation. Their ambitions and interests in the Arctic can thus tell us a lot about the Arctic Council's room for action. In Chap. 3, the Arctic Council's role in "Arctic governance" is discussed: Where do the political and legal constraints lie? In extension of this, the road ahead for Arctic collaboration is discussed (Chap. 4). Here, I dare to reflect more widely about what should be done to make sure the Arctic Council remains the most important international forum for Arctic-related dialogue and knowledge development. In the final chapter (Chap. 5), I recapitulate the main argument in this book and ask why we have an Arctic Council in the first place. By then, I hope the reader will have gained some understanding of the Arctic Council.

* * *

Throughout the Cold War, the characteristic feature of the Arctic was the military and strategic rivalry between the US and the Soviet Union. The Cold War stifled opportunities for collaboration in the region. The Arctic was a place for security policy and strategic assessments. All aspects of politics in the north were seen in light of this tense but essentially stable situation. In the 1980s, the thaw got under way. In his celebrated Murmansk speech of 1987, Mikhail Gorbachev was eager to transform the Arctic into a "zone of peace" and the North Pole into a "pole of peace".[3] Gorbachev's olive branch heralded new times. Canada and the US were both at the time determined to let the Arctic become a region of international cooperation.[4] By the end of the 1980s, the conditions were right for the Arctic states—including Russia—to create meeting places where common challenges facing the area could be discussed. This was nothing new, however. Norway and Russia had long been working closely together on the management of fish stocks in the Barents Sea,[5] for example. Now, though, all eight Arctic states could get together to discuss broader issues, not just narrowly defined fields.

In parallel with this development, environmental policy had climbed up the international agenda. The 1972 Stockholm Conference is often regarded as the start of international environmental politics. It resulted in, among other things, several international environmental agreements on, for example, conservation of endangered species. During the 1970s, several treaties aimed at combating pollution were signed. In the 1980s, fighting pollution became a global policy.[6] Steps taken to reduce ozone-depleting emissions, among other things, were important here. In 1982, the United Nations Convention on the Law of the Sea was signed. Marine pollution and management were key issues here. The 1992 Rio Conference on Environment and Development is considered a watershed event in international environmental politics. It was during this period the UN-appointed World Commission on Environment and Development launched the versatile concept of sustainable development,[7] a term that subsequently would play an important role in the debate on developments in the Arctic.

This positive atmosphere gave Finland the impetus in January 1989 to invite the Arctic states to come together to discuss the protection of the Arctic environment, under the aegis of the so-called Rovaniemi process.[8] The first meeting was held in Rovaniemi in September 1989 and the project was called the Arctic Environmental Protection Strategy (AEPS). In the ensuing years, the eight circumpolar states—the US, Russia, Canada, Sweden, Finland, Denmark (including Greenland and Faroe Islands), Norway and Iceland—met regularly. Working groups were established and indigenous groups and non-Arctic states were invited as well. In

June 1991, AEPS was officially established. An important arena for international cooperation saw the light of day.[9]

Alongside the rollout of the Finnish initiative, Canada was pushing an idea for an Arctic Council. The indigenous voice was particularly important.[10] A new forum for the Arctic states was first aired by Canada's then Prime Minister, Brian Mulroney, in Leningrad in November 1989. The idea was not to incorporate AEPS as part of the Arctic Council. The Canadians wanted rather to support the demilitarisation process that had taken hold since the end of the Cold War. Canada wanted this Arctic council to be a forum where states could discuss security issues, an idea welcomed by Finland. Nevertheless, there was resistance; the US in particular was sceptical. For the Americans, it was (and remains) important that such a forum should not address questions of security.[11] There are several reasons for this reluctance, but America's traditional mistrust of binding multilateral bodies addressing sensitive political questions is one.[12] Another controversy regarding the establishment of the Arctic Council centred on whether the forum should be expanded into an international organisation. The Arctic states agreed to the organisation of the Council as a forum,[13] and again, it was Americans who held back from committing themselves more than necessary. Such a forum has a weaker mandate than an international organisation and, among other things, has less legal clout (see Chaps. 2 and 3). In 1996, the Ottawa Declaration officially established the Arctic Council. The first sentence of the Declaration reads: "The Arctic Council is established as a high level forum to: (a) provide means for promoting cooperation, coordination and interaction among the Arctic States, with the involvement of the Arctic indigenous communities and other Arctic inhabitants on common Arctic issues." At the birth of the Arctic Council, AEPS was incorporated into the new forum, and indigenous peoples, observers, member states and working groups involved in AEPS came along too. Although participants considered the Council as a breakthrough for an official and constructive form of collaboration in the north, not all were equally enthusiastic. In the *Washington Post*, the achievement was relegated to an inconspicuous place deep inside the paper and described as an "international panel" designed to "protect native people [and] wildlife".[14] Not overwhelmingly effusive, in other words.

Procedures and rules were adopted at the first meetings. It was decided, among other things, to let the chairmanship rotate every other year among the eight member states and that all decisions should be taken unanimously. The role of observers and the status of indigenous peoples as permanent participants became clearer as well. The permanent participants have full consultation rights in connection with the Council's negotiations and decisions, but not voting rights. The indigenous groups that constituted the permanent participants stand for quite different interests and face different challenges. Nevertheless, in the Council's infancy, it was important to anchor the work in a traditional Arctic identity. Mary Simon in particular, then president of the Inuit Circumpolar Conference, was at the heart of setting up the Arctic Council.[15] The observers were and remain a highly diverse group of NGOs, international organisations and non-Arctic states. As the name suggests, they are meant to observe the work of the Council and primarily interface with the working groups. They don't have voting rights but shall be invited to the meetings of the Arctic Council. This organisation with eight permanent members, a rotating chairmanship, unanimity, permanent participants and a large number of observers characterises today's Arctic Council as well.[16]

* * *

In 1996, 30 projects were under way, and interest in the Arctic within the international community was limited. By 2019, the project number will have risen to about 100, and the number of actors with ambitions for the region will have grown exponentially. How did we get here and what role did the Council play? In the first period, from 1996 to the mid-2000s, most of the work of the Arctic Council centred on battling pollution in the Arctic. There are high levels of pollutants and heavy metals in the north, with the potential to harm to both wildlife and people in the Arctic. Much of this pollution is carried from the south to the north on ocean and air currents. It was not a particularly sensitive political question and was typical of its time when a range of environmental challenges were the subject of international negotiations. Environmental issues were highly relevant politically. This was evident in the work to monitor environmental pollution in the Arctic. One example is the Arctic Monitoring and Assessment Programme (AMAP), one of the most important working groups in the Council, which in its first major report of 1998 identified the prevalence of pollution in the Arctic.[17]

The Arctic was still "remote", and specific environmental issues loomed large on the Arctic agenda. During this period, the working groups prepared a series of reports describing these challenges. This work has gone on to inform international conventions on pollution, including the Minimata Convention (on mercury) and the Convention on Persistent Organic Pollutants (Stockholm Convention). Much of the Arctic Council's focus is still on these issues.

Early in the 2000s, awareness grew of climate change in the Arctic. This was evident among other places in the most widely discussed scientific work released by the Arctic Council, the Arctic Climate Impact Assessment (ACIA Report).[18] One could say that the ACIA brought the Arctic Council out of the valley of shadows into the global spotlight. The idea of ACIA was launched in the mid-1990s and work began around 2000.[19] The report showed the special situation in which Arctic finds itself regarding climate change. The rise in temperatures in the Arctic has been twice as fast as the global average, causing both sea and land ice to melt. The main point is that it is in the Arctic we will see the first and most significant signs of a changing climate. The impact of these developments on the local population has also received considerable attention. Changing ice conditions, erosion and less permafrost are changing the basic parameters. People's living conditions in the north will therefore change completely. At the same time, acidification of the seas and increased temperatures can change the delicate ecological balance of the Arctic. The indigenous peoples were entitled to an opportunity to participate actively in the work to prepare the ACIA report in light of their status as permanent participants. Their local knowledge was and is still often seen as crucial to understanding and tackling the major changes we are witnessing. In this context, the Council has addressed both emission reductions and adaptation to a changing climate. With regard to emission limits, particular challenges regarding the so-called short-lived greenhouse gases, such as soot and gases such as methane and ground-level ozone have been on the agenda.[20] In 2009, a time-limited task force was also appointed to work specifically on this topic,[21] an effort that has been of key significance since that time. Through ACIA and subsequent work on both adaptation to and emission limits, the working groups have provided input to international climate negotiations, especially through the Climate Panel (IPCC), and the Arctic has stood as an important symbol of global climate change.[22]

The main focus in recent years has been on adaptation to a changed climate, or more specifically, how to manage the increasing (and potentially escalating) commercial activity, which is partly made possible by the retreating sea ice. Two agreements negotiated by the eight member states on search and rescue (2011) and oil spill preparedness (2013) bear witness to this trend.[23] It is nevertheless necessary to point out that Arctic Council did not draft the agreements which were negotiated by the eight members states. On the other side, that the Council has served as a venue for the drafting of legally binding international agreements also betokens a more proactive role.[24] It could be a sign of more binding Arctic cooperation. These two agreements can stand as symbols of a new age, and therefore deserve particular attention.

* * *

At a meeting in Ilulissat in Greenland in 2008, the five states bordering the Arctic Ocean declared: "The increased use of Arctic waters for shipping, tourism, resource development also increases the risk of accidents and therefore the need to further strengthen the search and rescue capabilities and capacity around the Arctic Ocean." The Ilulissat Declaration further states that such a strengthening must cover "bilateral and multilateral arrangements between or among relevant states".[25] In consequence of this statement, search and rescue efforts in the Arctic moved higher up the agenda. However, a search and rescue agreement was not negotiated between the Arctic coastal states despite the fact the crucial importance to these states of an agreement on search and rescue in the Arctic. It is also important to note that this was not an Arctic Council initiative and only the coastal states were participating. On the other hand, the subject had been on the Council's agenda on previous occasions. In 2001, Russia proposed a preparatory project to be undertaken at the Arctic Council.[26] At a 2001 meeting of government officials, Russia expressed a wish to promote international cooperation in the event of accidents in the Arctic.[27] Russia introduced several initiatives in this regard and in 2004 the expectation that this work would result in an international agreement was officially noted for the first time. But it was only in Tromsø in 2009 that member states managed to agree to negotiate a search and rescue agreement for the Arctic. The US and Russia headed a time-limited task force entrusted with formulating such an agreement. The agreement was signed in Nuuk in May 2011.[28] Its main objective is formulated as follows in Article 2: "To strengthen aeronautical and maritime search and rescue cooperation and coordination in the Arctic." The agreement also clarifies the scope of the geographical areas for which the respective states are responsible. That the Arctic states could come together in support of such a document shows that the Arctic Council is considered a relevant arena to discuss the coordination of the extensive preparedness arrangements in the area.

At the ministerial meeting in Kiruna in 2013 an agreement on oil spill preparedness in the Arctic was signed. The agreement is largely patterned on the search and rescue agreement. It defines the respective states' areas of responsibility, cooperation being the key word. The work on oil spill preparedness also has a long history in the Arctic Council. The extraction and transport of oil and gas represent one of the biggest threats to the Arctic environment, as was also emphasised in the wide-ranging 1998 report on pollution.[29] Increasing activity, according to the Arctic Council, will only make this issue more relevant.[30] These efforts have led to the development of non-binding guidelines for oil and gas-related activity (including shipping) in the region.[31] In the 2009 Arctic Maritime Shipping Assessment (AMSA), better follow-up mechanisms are being developed when it comes to implementing policy decisions in the Arctic Council.[32] Here, then, the Council has tried to be a step ahead.

One of the driving forces behind the increase in interest in the Arctic is the undiscovered petroleum resources in the region. In 2008, the United States Geological Survey presented figures indicating that 13% of undiscovered oil resources and 31% of undiscovered gas resources in the world could be located in Arctic.[33] As mentioned, the Arctic Council has, over a long period of time, had a political and scientific focus on resource extraction, the potential increase in the rate of extraction and the environmental impact of such activity.[34] It is especially in the working groups this work has taken place. However, it is only in recent years that the issues have topped the agenda. In 2011, a task force was appointed to draft an oil spill preparedness plan.[35] The US, Russia and Norway led this work. At the ministerial meeting in Kiruna two years later, the agreement was signed.[36] Article 1 of the agreement states: "The objective of this Agreement is to strengthen cooperation, coordination and mutual assistance among the parties on oil pollution preparedness and response in the Arctic in order to protect the marine environment from pollution by oil." The purpose of the agreements negotiated by the Arctic states is not to set standards or specify capacity levels, but coordinate joint efforts through information exchange and multilateral exercises. How successful this work has been is another question (see Chap. 3). All the same, the agreements do testify to the wish of the eight Arctic states to strengthen Arctic cooperation.

* * *

In connection with the ministerial meeting in Kiruna in 2013, the issue of observer status for the EU and China was the most important item on the agenda and received most attention in international media. China and a range of other countries were awarded observer status after intense lobbying from, among others, US Secretary of State John Kerry and Sweden's Foreign Minister Carl Bildt.[37] Other countries awarded observer status at the meeting in Kiruna were India, Italy, Japan, Singapore and South Korea. The EU was also awarded observer status, but Canada and the EU had to agree on further conditions before their status could be finally approved.[38] The tense relationship between Canada and the EU on the matter of seal hunting was the main reason.[39] The EU's boycott of seal products was a heavy blow to Canada.[40] Either way, the desire for observer status of EU and China tells us that the region is considered important also by actors located far from the geographically delimited Arctic. The Arctic and Arctic Council are popular and geopolitically relevant. We are moving towards a new phase in the Council's history, where the core debate will centre on questions about the depth and breadth of cooperation.[41] How many observers can and should be allowed and what role can they play, and will the Council continue to grow as an arena for the formulation of legally binding international agreements? These are issues to which I return later in the book.

* * *

Concurrently, the economic development of the Arctic and the local population's participation in this upswing have attracted increasing attention. In this context, several have chosen to apply the variegated concept of sustainable development. This work, among others, was top of the agenda during the Canadian chairmanship (2013–2015). When the Canadians took over the chairmanship of the Arctic Council in 2013, their stated objective was "development for the people of the North".[42] During this period, an economic council for the Arctic was also established. With the creation of what became the Arctic Economic Council (AEC), the Arctic states showed their wish to give these questions greater weight. The AEC was a Canadian initiative, although its secretariat is based in Tromsø. This newly established forum has taken quite a long time to settle in, and how it is meant to relate to Arctic Council remains unclear.

At the same time, the Canadian chairmanship period was characterised by international turbulence. The Ukraine conflict had an impact on Arctic collaboration. When the latter was discussed, one could hear statements such as: "In the Arctic Council, we go from crisis to crisis."[43] Most states, however, have been keen to shield the Arctic Council from geopolitical "noise". Work in the working groups has progressed further. In 2015, the US took over the chairmanship of the Council, and again it was emphasised that conflicts in Ukraine should not be allowed to disrupt the important work of the Council. During the US chairmanship, the economic development of the Arctic community was also high on the agenda, but climate change and environmental challenges in the north were also attracting more attention. During the Finnish chairmanship, which started in 2017, technological development in the north is a key point. In recent years, states have thus expressed a marked interest in integrating Arctic communities in the global economy during various chairmanship periods. The Arctic is no longer "remote". One last item to set its stamp on the wider Arctic debate, but on the Arctic Council itself, is the balance between growth and protection. To what extent should the vulnerable Arctic environment be protected through, for example, the creation of protected areas? Not surprisingly, the coastal states want to extract and manage the resources found in the region without too much interference from states that have neither sovereign rights nor jurisdiction in the area. This discussion has not reached its conclusion.

* * *

The Arctic Council has thus grown from a limited forum espousing environmental cooperation to extensive collaboration across a wide range of Arctic challenges. This could be explained simply by the fact that institutions tend to expand over time, and especially after successfully completing work. At the same time, the world's eyes have increasingly targeted the region. The Arctic, in the opinion of some, is at the geopolitical centre. In recent years and on several occasions, we have been able to read that the path to a major conflict in the Arctic is not necessarily long. Such views were especially prominent in the aftermath of the exploits of a Russian polar research delegation led by Artur Chilingarov which, using three-person Mir-1 mini submarine, planted a titanium flag underneath the North Pole in 2007. One of the most quoted contributions in this regard was penned by former US Coast Guard officer Scott Borgerson. In a 2008 article in *Foreign Affairs* he warned, "[W]ithout U.S. leadership to help develop diplomatic solutions to competing claims and potential conflicts, the region could erupt in an armed mad dash for its resources."[44] Similar reasoning informed coverage in the international press. A 2007 article in *Time Magazine* stated, for example, "[J]ust as 150 years ago, when Russia and Britain fought for control of central Asia, it is tempting to think that—not on the steppe or dusty mountains but in the icy wastes of the frozen north—a new Great Game is afoot."[45] The picture of the Arctic has become less black and white in recent years, but that the conflict potential remains high is still an idea held by journalists and others. The question is whether these notions are only overdramatised representations of the new reality that is slowly taking shape in the Arctic.

In the now widely discussed future of the Arctic, we can distinguish between analyses and political comments based on a view of the international system as essentially anarchic, lacking a supreme authority where might is right, on the one hand, and those who prefer to put the spotlight on different collaborative mechanisms, on the other. When vital interests are under pressure, some argue, states will resort to force to make clear their positions. Scott Borgerson's analysis can be located here. We can also find book titles like *Arktisches Monopoly*, *The Arctic Gold Rush*, *The Scramble for the Arctic* and *The Cold Front*.[46] Having said that, we find a number of contributions highlighting the potential for cooperation. The majority of analyses written by researchers who have been working on Arctic questions over a long period of time belong to this second category. In the wake of the Russian's flag planting exploit, researcher Torbjørn Pedersen said: "The 'race' for areas of the shelf is proceeding in an orderly fashion in compliance with criteria outlined by the Convention on the Law of the Sea."[47] He attacked what he called the myths about the Arctic. Professor Rolf Tamnes at the Norwegian Institute for Defence Studies said the chances of a military confrontation are small: "There is a limited potential for conflicts. The region has a long history of peaceful cooperation."[48] Professor Lawson Brigham applied the same logic in a 2010 article in *Foreign Policy*: "The world has plenty of regions where serious conflict is a way of life already. Let's worry about them first." As mentioned, these somewhat more dispassionate arguments have begun to gain a foothold.[49] Ideas of the Arctic as perching on the threshold of chaos are, however, still prominent in the media. They are today being more closely linked to the strained relationship between Russia and the West, especially in the aftermath of the crisis in Ukraine and Russia's annexation of the Crimean Peninsula.

There are especially three arguments that back the hypothesis that the chances of civilised conflict resolution are greater than the risk of a rapidly escalating use of force. First, unclear and ambiguous projections on climate change, both regional and global, make planning difficult. Second, and in extension of the first, it is uncertain when and how resources in the Arctic will be exploited. And anyway, most of these resources are located within clearly demarcated boundaries. Further, the image of the Arctic Ocean as a coming competitive shipping route is rather exaggerated even though regional shipping and tourism are increasing. Third, we see a generally positive climate of cooperation. The Arctic states have acknowledged that sustainable management and economically viable resource exploitation in the region require international cooperation. They have a common interest in the region remaining stable. In conclusion: the potential for a civilised Arctic also in the future is therefore significant.

At the same time, Arctic cooperation does not exist in isolation from the Realpolitik reality. Russia's actions in Ukraine have, for example, had an impact on the collaboration. The Canadians and Americans boycotted, among other things, an Arctic Council meeting in Moscow in the autumn of 2014,[50] and the Canadians have on occasion used rather powerful terms against Russia, also in forums where Arctic cooperation was on the agenda. This was a contributing reason why Russia's Foreign Minister Sergei Lavrov did not attend the Arctic Council's Ministerial Council in Canada in 2015.[51] The Arctic Council's options are thus shaped by wider international processes and the way states respond to them. But the main picture is that despite the cooling down of relations between Russia and the West after the annexation of the Crimea, the work of the Arctic Council continues.

The danger of the tabloid portrayals of the Arctic is the risk of ending up in a situation where all questions about the future in the north are mixed together, and where one concludes that the road to conflict is short. Such repeated prophecies are regrettable. Portraying the Arctic as an uncontrolled region is not just wrong, it should not be accepted in the public debate. It drowns out other Arctic debates that are more important, such as those on climate change, socio-economic development and the environment. And this is where the Arctic Council comes in.

A more nuanced depiction of how the Arctic Council works will make it easier to see what is right and important to look at when the development of the Arctic is the subject. The forum is perhaps the most important place to discuss the challenges facing us in a changing Arctic. A related question is nevertheless where the limits of the Arctic Council actually go. What can the Arctic Council do, and in extension of this and at least as relevant, what should the Arctic Council do to uphold its reputation as the most important international forum for cooperation on opportunities and challenges in the Arctic? In order to discuss these questions, it is first necessary to look into how the Council is organised and what the member states say about its significance.

Notes

1. Rethinking the Top of the World: Arctic Public Opinion Survey, Vol. Published by Munk-Gordon Arctic Security Program, April 2015. Available at http://gordonfoundation.ca/app/uploads/2017/03/APO_Survey_Volume-2_WEB.pdf. Last visited 24 April 2019.
2. Although we find previous books on the Arctic Council (e.g. John English: *Ice and Water. Politics, Peoples, and The Arctic Council* (Toronto: Allen Lane, 2013) and Douglas Nord: *The Arctic Council: Governance Within the Far North*) this book is different. In the sense that is focuses more explicit on what the Arctic Council is and what it is not by devoting much space on the work of the working groups and discussing the Council's actual role in Arctic governance. This book is the result of research undertaken in connection with the POLGOV project (no. 257664), funded by the Research Council of Norway.
3. Mikhail Gorbachev (1987) Speech in Murmansk, 1 October1987. Available at https://www.barentsinfo.fi/docs/Gorbachev_speech.pdf. Last visited 24 April 2019. Gorbachev was critical of Western policy in the region, but it's the positive message that's remembered today.
4. Torbjørn Pedersen (2012): "Debates over the Role of the Arctic Council". *Ocean Development & International Law*, 43(2), pp. 146–156, p. 147.
5. Geir Hønneland: *Kvotekamp og kyststatssolidaritet: Norsk-russisk fiskeriforvaltning gjennom 30 år* (Bergen: Fagbokforlaget, 2006).
6. Steinar Andresen, Elin Lerum Boasson and Geir Hønneland (eds): *International Environmental Agreements: An Introduction* (London/New York: Routledge, 2012).
7. The concept was central to the UN publication Report of the World Commission on Environment and Development: Our Common Future (1987). Available at http://www.un-documents.net/wced-ocf.htm. Last visited 24 April 2019.
8. John English: *Ice and Water. Politics, Peoples, and The Arctic Council* (Toronto: Allen Lane, 2013).
9. For a detailed description of this process and the political and diplomatic play around the establishment of AEPS see ibid.

10. English: *Ice and Water. Politics, Peoples, and The Arctic Council*, p. 172.
11. In the Ottawa Declaration of 1996 it is clearly stated that the Arctic Council should not address security issues, see https://oaarchive.arctic-council.org/handle/11374/85. Last visited 24 April 2019. For a wider discussion of the issue, see Chap. 3.
12. English: *Ice and Water. Politics, Peoples, and The Arctic Council.*
13. The Arctic Council is a forum without "legal personality" and is thus not an international organisation, strictly speaking. For a more detailed account of this distinction, see Morten Ruud and Geir Ulfstein: *Innføring i folkerett* (Oslo: Universitetsforlaget, 2002) pp. 81–84 and Evan T. Bloom (1999): "Establishment of the Arctic Council". *The American Journal of International Law*, 93 (3), 712–722. Another example of this type of international forum is the OSCE.
14. English: *Ice and Water. Politics, Peoples, and The Arctic Council*, p. 175.
15. Ibid., pp. 171–180.
16. For a more detailed presentation of the structure and organisation of the Council, see Chap. 2.
17. AMAP (1998), *AMAP Assessment Report: Arctic Pollution Issues.* Available at https://www.amap.no/documents/doc/amap-assessment-report-arctic-pollution-issues/68. Last visited 24 April 2019.
18. ACIA (2005) *Arctic Climate Impact Assessment.* Available at https://www.amap.no/arctic-climate-impact-assessment-acia. Last visited 24 April 2019.
19. For a detailed analysis of ACIA's history and significance, see Timo Koivurova (2008): "Transboundary Environmental Assessment in the Arctic", *Impact Assessment and Project Appraisal*, 26 (4), pp. 265–275.
20. Short-lived climate forcers (SLCFs) are gases and particles that contribute to heating and have a lifespan between a few days and up to 15 years.
21. Task Force on Short-Lived Climate Forcers (SLCFs), See https://arctic-council.org/index.php/en/task-forces/76-slcf. Last visited 24 April 2019.
22. For a review of the Arctic Council's role in this regard, see Sebastian Duyck (2012) "Which Canary in the Coalmine? The Arctic in the International Climate Change Regime" in Timo Koivurova, Gudmundur Alfredsson and W. Hasanat, *The Yearbook of Polar Law*, Vol. 4. Leiden: Bill Editions: 583–617.
23. A third agreement on research collaboration was signed at the ministerial meeting in Fairbanks in 2017.
24. Svein Vigeland Rottem (2013) "The Arctic Council and the Search and Rescue Agreement: The case of Norway". *Polar Record*, 46 (2), pp. 146–156.

25. Ilulissat Declaration (2008). Available at https://cil.nus.edu.sg/wp-content/uploads/2017/07/2008-Ilulissat-Declaration.pdf. Last visited 24 April 2019.
26. For analyses of Russia's role in the Arctic Council in general and in search and rescue work in particular, see Ingmar Oldberg (2011): "Soft Security in the Arctic". *Occasional UIpapers, Published by Swedish Institute of International Affairs*, No. 4; Roald Ramsdal (2012): *Russland i Arktisk Råd*. Master's thesis, UiO; Elana Wilson Rowe (2008): "Russian Regional Multilateralism: The Case of the Arctic Council" in Elena Wilson Rowe and Solveig Torjesen (eds) *The Multilateral Dimension in Russian Foreign Policy*, Abingdon: Routledge: 142–152.
27. Ramsdal (2012): *Russland i Arktisk Råd*, p. 58.
28. The full name of the agreement is *Agreement on Aeronautical and Maritime Search and Rescue in the Arctic*. Available from https://oaarchive.arctic-council.org/handle/11374/531. Last visited 24 April 2019. For a closer analysis of the content of the agreement, see Shih-Ming Kao, Nathaniel S. Pearre and James Firestone (2012) "Adoption of the Arctic Search and Rescue Agreement": *Marine Policy*, 36 (3) 832–838, pp. and Rottem (2013) "The Arctic Council and the Search and Rescue Agreement: The case of Norway". The agreement entered into force on 19 January 2013.
29. AMAP (1998) *AMAP Assessment Report: Arctic Pollution Issues*. Available from https://www.amap.no/documents/doc/amap-assessment-report-arctic-pollution-issues/68. Last visited 24 April 2019.
30. AMSA (2009) *Arctic Maritime Shipping Assessment*. Available from https://www.pame.is/index.php/projects/arctic-marine-shipping/amsa. Last visited 24 April 2019.
31. See, for example, PAME (2002) Arctic Council. Arctic Offshore Oil & Gas Guidelines. Available from https://oaarchive.arctic-council.org/bitstream/handle/11374/1591/MM03_PAME_Attachment_4.pdf?sequence=5&isAllowed=y. Last visited 24 April 2019.
32. For an analysis of the national implementation of AMSA, see Ida Folkestad Soltvedt (2017): "Soft Law, Solid Implementation? The Influence of Precision, Monitoring and Stakeholder Involvement on Norwegian Implementation of Arctic Council Recommendations". *Arctic Review on Law and Politics*, Vol. 8, pp. 73–94.
33. USGS 2008 World Petroleum Assessment. Available from https://energy.usgs.gov/OilGas/AssessmentsData/WorldPetroleumAssessment.aspx. Last visited 24 April 2019.
34. AMAP (2007) *Arctic Oil and Gas*. Available from https://www.amap.no/oil-and-gas-assessment-oga. Last visited 24 April 2019.

35. See Arctic Council, Task Force on Arctic Marine Oil Pollution Preparedness and Response, https://arctic-council.org/index.php/en/task-forces/72-tf-marine-oil-pollution-preparedness-response. Last visited 24 April 2019.
36. The full name of the agreement is "*Agreement on Cooperation on Marine Oil Pollution Preparedness and Response in the Arctic*". Accessible from https://oaarchive.arctic-council.org/handle/11374/529. Last visited 24 April 2019.
37. Philip E. Steinberg and Klaus Dodds (2015) "The Arctic Council after Kiruna". *Polar Record*, 51(1), pp. 108–110.
38. At the time of writing, the issue is still not clarified. Nor was it possible to achieve consensus at the 2015 ministerial meeting in Canada. States chose rather to postpone the question of observer status for several countries. Not surprisingly, it was an extremely sensitive political issues in light of the crisis in Ukraine, and it strained the relationship between Russia and the other Arctic states as well as the EU. At the 2017 ministerial meeting in Fairbanks, Alaska, the EU was not officially invited to attend either despite the fact that Switzerland, among others, had been awarded observer status. For all practical purposes, however, the EU is considered an observer.
39. For an analysis of this conflict, see Njord Wegge (2013) "Politics between science, law and sentiments: explaining the European Union's ban on trade in seal products", *Environmental Politics*, 22 (2), pp. 255–273.
40. Andrew Chater argues that opposition from Inuit Circumpolar Council (ICC) forced Canada to block accredited observer status for the EU. See Chater, Andrew (2016): "Explaining Non-State Actors in the Arctic Council". *Strategic Analysis*, 40(3), 173–184.
41. See Axworthy, Koivurova and Hasanat *The Yearbook of Polar Law*; and Lassi Heinien, Heather Exner-Pirot and Joël Plouffe (2016) *The Arctic Council: 20 Years of Regional Cooperation and Policy-Shaping, Arctic Yearbook* for more contributions to the discussion of the future of the Arctic Council.
42. See the Canadian Chairmanship Program. Available from https://www.arctic-council.org/index.php/en/our-work2/8-news-and-events/226-canadian-chairmanship-program-2013-2015. Last visited 24 April 2019.
43. Informal conversation with a representative of the Arctic Council, 14 April 2014.
44. Scott G. Borgerson, 'Arctic Meltdown', *Foreign Affairs*, 87.2 (2008), 63–77.
45. James Graff. 'Fight for the top of the world'. *Time Magazine* 19 September 2007. Available from http://content.time.com/time/magazine/article/0,9171,1663848,00.html. Last visited 24 April 2019.
46. Several books can stand for this approach to Arctic questions. See, for example, David Fairhall: *Cold Front: Conflict Ahead in Arctic Waters*

(London: I.B. Tauris, 2010); Robert Howard: *The Arctic Gold Rush* (London: Continuum, 2009); Richard Sale and Eugene Potapov: *The Scramble for the Arctic* (London: Frances Lincoln Ltd., 2010); Christoph Seidler: *Arktisches Monopoly* (DeutscheVerlags-Anstalt, 2009). The various contributions are more or less conflict-oriented, but the titles still say something about how the Arctic is portrayed. For an analysis of these discourses see, for example, Øyvind Østerud and Geir Hønneland: "Geopolitics and International Governance in the Arctic"; Svein Vigeland Rottem and Ida Folkestad Soltvedt (eds), *Arctic Governance: Law and Politics. Volume 1.* (London: I.B. Tauris, 2017).

47. Torbjørn Pedersen: "Utenriksanalyse: Knus mytene om Arktis". *Morgenbladet* 24 July 2009.
48. Rolf Tamnes. Quoted in *Teknisk Ukeblad* in the article, "Det arktiske monopolspillet". 6 October 2009. Available at https://www.tu.no/artikler/det-arktiske-monopolspillet/256007. Last visited 24 April 2019.
49. Lawson Brigham: "Think again: The Arctic". *Foreign Policy.* 6 August 2010.
50. CBS News. "Canada boycotts Arctic Council Moscow meeting over Ukraine". 16 April 2014. Available at https://www.cbc.ca/news/canada/north/canada-boycotts-arctic-council-moscow-meeting-over-ukraine-1.2611964. Last visited 24 April 2019.
51. CBS News. "Russia's foreign minister passes on Iqaluit Arctic Council meeting" 14 April 2015. Available from https://www.cbc.ca/news/canada/north/russia-s-foreign-minister-passes-on-iqaluit-arctic-council-meeting-1.3031430. Last visited 24 April 2019.

CHAPTER 2

Organisation, Actors and Member States' Interests

Abstract This chapter outlines the three levels of work in the Council: ministerial level, the senior Arctic official (SAO) level and working group level. It also introduces the reader to the work of Council by outlining key accomplishments by the six working groups and provides an overview on the role of the observers and the permanent participants (indigenous groups). The chapter also discusses the attitude of the member states to Arctic developments and the Arctic Council and argues that the Arctic Council has created an arena and network of different players (observers, indigenous groups and scientists) that the member states cannot ignore when the future of the Arctic is discussed.

Keywords Member states • Observers • Permanent participants • Network

The statutes governing the Council were adopted at the 1998 meeting of ministers at Iqaluit, Canada.[1] A key element of these statutes is that all decisions in the forum and underlying working groups require all eight permanent member states to agree.[2] Furthermore, only member states and permanent participants (indigenous peoples' groups) can formally submit proposals for new projects. It was also decided that the chairmanship should rotate between the eight permanent members. A chairmanship period generally lasts for two years. The main work of the Council

S. V. Rottem, *The Arctic Council*,
https://doi.org/10.1007/978-981-13-9290-0_2

itself takes place at three levels: the ministerial level, the SAO level and working group level. All are supported by the secretariat in Tromsø.[3] The work of the secretariat rests on two main pillars: administration and communication. The secretariat is therefore more of a facilitator than an active contributor, for example, to setting a political agenda.

When the Arctic States hold ministerial meetings, the Arctic Council does tend to attract the public spotlight. Statements from the ministerial meetings tell us how the member states would like the Council to develop. This is where the fundamental direction can be made out. While member states used not to be represented by their foreign ministers at these meetings in the infancy of the Council, there has been a change in recent years. For instance, the 2011 ministerial meeting in Nuuk in Greenland was attended by Russian Foreign Minister Sergei Lavrov and US Secretary of State Hillary Clinton. At the 2013 Kiruna meeting, the US secretary of state and Russian foreign minister (John Kerry and Sergei Lavrov) added lustre to the gathering. As previously mentioned, while Lavrov did not attend the 2015 meeting, he was back for the 2017 meeting together with the newly appointed Secretary of State Rex Tillerson from the US. The Council has arguably grown in importance as a multilateral political forum (but also as a platform for launching binding agreements negotiated by the Arctic States). The participation of US secretaries of state and Russian foreign ministers gives greater political urgency to the communications from the north. In connection with the ministerial meetings, the States also hand over chairmanship responsibilities. In spite of its formal position as the highest decision-making body, the ministerial meeting has traditionally played more of a symbolic role, applauding what is being done and agreeing to follow the same path in the future too. But at these ministerial meetings, the State holding the chairmanship makes its own ambitions known, and here member states have an opportunity to put their own personal stamp on proceedings. The actual work, however, takes place further down the organisational ladder.

* * *

At the level of government officials, the so-called SAOs meet at least twice a year. Each State appoints an SAO to advance their interests in the Arctic Council. The SAOs thus represent their government, and are usually nominated from among the State's foreign ministry officials. Their formal duty is to guide and oversee the Arctic Council's activities in accordance with the decisions and instructions of the Arctic Council's foreign ministers. They act, in other words, as a link between the ministerial level and the working groups. It is their job to follow up the work of the Council.[4] The fact that the SAOs traditionally hail from their respective State's foreign ministries and are familiar with the ways of diplomacy, procedures and of building relations rather than technical details requiring specialist knowledge of the Arctic is a challenge. Where to strike the balance between politics and science is therefore a recurring topic. Another challenge is the more or less regular replacement of the SAOs.[5] SAOs and ministers come and go while the working groups remain in place. Continuity is a problem and can make it difficult to achieve a satisfactory understanding of what is possible and sensible. There are several reasons for this unending turnover of diplomats. One is their extensive travel itinerary. All States seem to want to host an international conference on the Arctic, such as the Arctic Frontiers conference in Norway, the Arctic Circle conference in Iceland and Arctic Futures in Brussels. Put bluntly, the constant rolling out of new conferences could be named an Arctic Flying Circus. Participation at meetings and conferences can wear out the most travelled diplomat. There's another consequence of the Arctic hype. Traditionally, the SAO meetings were where the working groups sought approval for their priorities. Now, Arctic scientists in the working groups are afraid that political priorities are being given more weight and that research is being increasingly manipulated to benefit political goals (see below).[6] What we might call basic research is being undermined.

* * *

The heart of the Council itself is the working groups. They can be described as producers of scientific knowledge, charged with identifying and analysing Arctic challenges, and are the major venue in which permanents participants and observers participate in the work of the Council. This is where most of the Arctic Council's work takes place. It covers everything from, for example, the monitoring of mercury levels to formulating guidelines for Arctic shipping to undertaking wide-ranging studies of climate change in the Arctic. The groups are just as likely to submit recommendations on everything from the regulation of environmental pollutants to choices of general policy. The working groups have been called the Council's locomotive.[7] They have also been the subject of a number of studies.[8] In one of the most quoted studies, Paula Kankaanpää and Oran Young say the working groups have been successful in identifying new challenges in the Arctic.[9] What we lack are studies of policy implementation and the transfer of knowledge from the working groups to international and national decision-makers (see Chap. 4).

The Arctic Council has six working groups: Arctic Monitoring and Assessment Programme (AMAP); Conservation of Arctic Flora and Fauna (CAFF); PAME (Protection of the Arctic Marine Environment); Emergency Prevention, Preparedness and Response (EPPR); Arctic Contaminants Action Programme (ACAP); and Sustainable Development Working Group (SDWG). Each working group has its own history and portfolio. They also vary in size from the three "big" programmes, AMAP, CAFF and PAME, to the three "small" groups, EPPR, ACAP, SDWG. But they share some common features. All working groups have representatives from the State's sector administrations, national administration and the scientific community. They have a specific operating mandate; a chairmanship (which also rotates between the member states); and a board or steering committee supported by a secretariat. All decisions require consensus, as does the rest of the Council's work. It is also important to emphasise that the working groups obtain their mandate from the ministerial meetings and the SAOs. On the other hand, it is easier to reach agreement on a recommendation at the working group level than at the SAO or ministerial level. This does not mean that a recommendation is necessarily taken up by the SAO or ministerial level, but the knowledge generated and the recommendations formulated will nevertheless set the agenda and, at best, put pressure on those able to take political action. A brief presentation of the individual working groups is necessary to understand what the Arctic Council does on a daily basis. This is where most of the work in the Council takes place.

AMAP, which has its secretariat in Tromsø, is often called the most important working group. Its budget is bigger than all the other working groups and it has a large secretariat with a staff of about seven. The funding of various projects has also been relatively stable. AMAP was founded in 1991 under the AEPS and became part of the new Arctic Council structure when the Council was formed. The working group's main task is to monitor the Arctic environment, ecosystems and population groups, and feed scientific advice and recommendations to decision-makers on how to deal with pollution and the harmful effects of climate change. For example, AMAP together with CAFF (see below) led the preparation of the Arctic Climate Impact Assessment (ACIA). The report resulted in recommendations on how to tackle climate change in the Arctic; the work on adaptation measures proved especially important. In AMAP's infancy, as mentioned earlier, the main focus was on various sources of pollution in the Arctic, such as persistent organic pollutants (POPs), heavy metals and radio activity. Local sources of pollution like oil and gas were also on the agenda. What we have witnessed over the years is a continuous expansion of AMAP's areas of work. From a rather limited focus on pollution in the Arctic environment to climate change (both emissions reduction and adaptation), and an increasing focus on the socio-economic consequences of these developments, AMAP's progress matches the general trend in multilateral collaboration on Arctic issues. The working group has provided input to various international conventions. The United Nations Environment Programme (UNEP) on pollution, the global climate regime in general and the Intergovernmental Panel on Climate Change (IPCC) can be mentioned as particularly important.

The working group with the mandate closest to AMAP is CAFF. CAFF's secretariat is in Akureyri, Iceland. Like AMAP, CAFF was one of the original working groups created under AEPS. CAFF has nevertheless historically been less visible than AMAP and has also had fewer resources.[10] This notwithstanding, CAFF has been responsible for several significant scientific publications.[11] The main thrust of the group's work is directed at Arctic flora and fauna and the conservation of Arctic biodiversity. In the 1990s, its focus was largely on individual species and their habitats. This changed with the ACIA report, which marked the emergence of a broader understanding of ecosystems. This needs to be seen in light of the growing global interest in ecosystems and ecosystem services.[12] In parallel, CAFF shifted its focus from collaboration on administrative and political issues to scientific work through monitoring and assessment of the health of Arctic ecosystems. This work reached a preliminary peak with the publication of the Arctic Biodiversity Assessment (ABA), unveiled at the May 2013 meeting of ministers in Kiruna. ABA paints a broad picture of the status of Arctic biodiversity. We also find 17 recommendations on consolidating efforts to conserve biodiversity and ecosystems. The majority of these recommendations are directed at the future work of the Council. They tell us what the Council should be concentrating on and where the gaps in knowledge are. But there are also recommendations to formulate national guidelines and to strengthen international obligations.[13]

Protecting areas of the Arctic has been a prominent subject of discussion in the group. It was a major feature of CAFF's work as early as 1991. However, with the increased focus on scientific work from the mid-2000s, the topic receded into the background of CAFF's daily operations. In connection with the launch of ABA in 2013, the subject of protected areas regained its place on the agenda. The question was whether CAFF should play an active role in promoting a series of Arctic protected areas. This is a hot political potato because it touches on issues of sovereign rights.[14] Several Arctic States are reluctant to get involved in work that is likely to affect the management of areas where Coastal States have sovereign rights, for example by establishing a network of Arctic protected areas.[15] Coastal States refuse to allow the Arctic Council to influence national priorities. This is a subject to which I shall return in the discussion of what the Arctic Council could become looking ahead (Chap. 3). This scepticism is also linked to the question of how "political" the Arctic Council should and can be. The Council alternates between being a producer of basic science and a venue where more or less binding political (and sometimes legally binding) recommendations are formulated.

The third "big" working group is PAME. PAME has also focused in its work on issues to do with protected areas in the Arctic. Like CAFF and AMAP, PAME was established under AEPS, and became a working group in formal terms in 1993. PAME's main mandate is to work on the protection and sustainable use of the Arctic marine environment. The most important achievement in recent years is the Arctic Maritime Shipping Assessment (AMSA). Here, PAME made numerous recommendations and helped present the most comprehensive survey of the challenges created by and facing maritime transport in the Arctic. AMSA has also provided input to IMO (International Maritime Organization)[16] and contributed to efforts to negotiate binding regulations for shipping in the region through the so-called Polar Code.[17] Through AMSA, steps were taken to enable monitoring mechanisms on enacting decisions and recommendations the group has formulated. Such mechanisms are quite unique in an Arctic Council context, and when they were established in 2009, it was uncharted territory for the Council. The AMSA report was presented at the 2009 Tromsø meeting, marking the end of the Norwegian chairmanship. The report states: "Increased marine access and navigation in the Arctic Ocean call for development and implementation of suitable national and international regulations, where appropriate, to advance the safety of Arctic marine shipping, including marine pollution prevention, reduce accident risk and facilitate effective emergency response." Ambitious goals which may or may not prove easy to achieve. PAME is not, therefore, a technical working group monitoring the Arctic environment, but more a venue to discuss policy measures that can meet new challenges in the north such as increased shipping.

In the work on maritime safety, the EPPR working group also plays a central role. EPPR has its secretariat in Tromsø and is also one of the first working groups to be formed under the AEPS. EPPR's mandate is mainly to ensure the protection of the Arctic environment from acute discharges of, for example, oil. The group's most important areas of concern in recent years have been the two negotiated agreements on search and rescue and oil spill preparedness. The working group will, among other things, update the guidelines annexed to the agreement on oil spill preparedness. The follow-up of the two agreements was formalised in 2017 with the establishment of expert groups.[18] We can therefore say that this working group is oriented more towards practical action and that it lacks the monitoring function of AMAP and CAFF. The scientific assessment of the impact, of, for example, an acute discharge, is not part of its mandate.

The major difference between the working groups is also evident in ACAP. This working group shares a secretariat with EPPR in Tromsø. ACAP was originally a programme intended to follow up AMAP's work to identify sources of pollution. It didn't become an official working group under the Arctic Council until 2006. ACAP has mostly concentrated on pollution from Russian sources, the main objective being to accelerate the reduction of national emissions of various environmental pollutants and climate drivers. There are four expert groups under ACAP, all working on practical projects. ACAP can thus be described as an initiator of projects whose scientific basis lies in AMAP's assessments of the health of the Arctic. The idea behind the creation of ACAP was to channel the work of the Arctic Council in a more practical direction and to improve the use of the knowledge produced by AMAP. As with some of the other Arctic Council working groups, it is sometimes unclear how different projects have been evaluated, how well they are coordinated with the other working groups and what they have achieved. All other working groups have nevertheless been evaluated and analysed by external bodies. As for ACAP, we know less; it is in fact the working group we know least about.[19] There isn't the space here for a more detailed analysis of the working group, but the fact that we know little about ACAP says something about the Arctic Council more widely, a matter to which I return to Chaps. 3 and 4.

One last interesting construction is the SDWG with its small secretariat in Canada. This working group focuses on economic and social development in the Arctic. Perhaps the most important work undertaken under the aegis of the working group is the 2004 Arctic Human Development Report (AHDR),[20] a project that has been extended. In the winter of 2015, The Arctic Development Report II (AHDRII) was published. The theme was regional processes and global links. The reports provide independent scientific assessments of human and regional development in the Arctic. These documents have not been welcomed by all but are important as they represent some of the first reports to focus specifically on social and economic development in the region.[21] There is a large number of smaller projects as well, ranging from food culture to children's health and everything in between.[22] So while SDWG has undoubtedly carried out a number of projects, exactly how many and how successful they have been is not always clear.[23]

When the Arctic Council's working group structure is critically evaluated, it is usually the SDWG's rather unclear mandate that receives attention.[24] The group is not one of the original working groups and was established in 1998, precisely because there was a palpable need to formalise the work on sustainability. A recurring challenge is that while the large working groups tend to focus on pan-Arctic challenges—which most agree deserve attention—such as climate change, shipping and pollution, this working group often has a more limited and localised focus on indigenous peoples in the region. In this connection, it has been argued by commentators that sustainable development should be a key component of the mandate of all of the working groups. The distinction between environmental problems and sustainable development is, they say, essentially contrived.[25] The map is at odds with the terrain. This is an insight that will be addressed in more detail in Chap. 4.

* * *

After this brief portrayal of the working groups, what stands out are the overlapping commitments off the groups. For example, they all have a maritime focus and are working on how climate and environmental changes affect the living conditions of people in the north. The working groups have also joined forces several times in the publication of reports. The big climate report from 2005 (ACIA) is once again a good example, whose scientific basis was something to which CAFF and AMAP in particular contributed. On the other hand, these similarities have resulted in recurring discussions about the need to improve inter-group coordination and examine whether there are too many working groups with overlapping functions. If there are, some should be merged. Again, this is a discussion that will be taken up in Chap. 4.

We have also seen an increase lately in the use of Task Forces (TF), that is, specialised and temporary expert panels, set up at ministerial meetings to look at specific issues. Their investigations have covered everything from economic cooperation to steps to limit climate emissions in the region.[26] The most well known are the ones that led to the three international legally binding agreements negotiated under the aegis of Arctic Council: the Task Force on Search and Rescue; the Task Force on Arctic Marine Oil Pollution Preparedness and Response; and the Task Force for Enhancing Scientific Cooperation in the Arctic (SCTF). These three TFs have completed their work. At the time of writing, two groups are involved in two other projects. They are the Task Force on Arctic Marine Cooperation (TFAMC) and the Task Force on Improved Connectivity in the Arctic (TFICA). Such groups usually consist of experts from one or more of the six working groups and representatives of the member states and any other implicated stakeholders. This relatively new arrangement of appointing time-limited expert groups has been criticised by some for making it even less clear what Arctic Council exactly can do and should be doing. The working groups have been particularly sceptical to the establishment of ad hoc groups whose investigations may encroach on their own mandates.[27] A new structure in an attempt to strengthen political control, but what one risks is an unclear division of labour and poor coordination.[28]

That the relationship between what is thought of as the coordinating level (ministerial meetings, SAOs and secretariat) and the scientific level (working groups) does not always operate flawlessly should not come as a surprise to anyone with a little understanding of policy making. Politicians want scientists to march in a particular direction, which the scientists are not always comfortable about doing. The somewhat freewheeling role of the working groups has often been commented on[29]; for example, AMAP used to operate without using the Arctic Council's logo. Essentially, the working groups had been allowed to follow their own inclinations without too much political intervention. Some commentators are seeing a change here, however, not only with the establishment of task forces, but with the ministerial meetings laying down clearer guidelines for the work of the working groups.[30] This may indicate a greater determination on the part of member states to use Arctic Council for political ends, for example as a platform for drafting binding agreements. Another point of interest is that the projects and work of the Council are funded by the member states on a voluntary basis. This means, among other things, that the working groups have to seek funding from a variety of sources, and member states can be quite selective regarding the projects they are willing to fund. At the same time, member states have to approve the projects before the working groups can start on them. The Arctic Council does not have a programme budget of its own.[31] The attitude of the member states to new and older projects is thus the driving (or counter) force in the cooperation, a factor that has become clearer since political interest in the north gathered pace.

* * *

Of the four million people living in Arctic areas, 500,000 belong to indigenous populations. They are thus an important voice in the Arctic alliance. Six international indigenous peoples' organisations enjoy status as permanent participants in the Council and they have as previously mentioned full consultation rights in connection with the Council's negotiations and decisions.[32] They are the Aleut International Association; the Arctic Athabaskan Council; the Gwich'in Council International; the Inui Circumpolar Council; the Russian Association of Indigenous Peoples of the North; and the Sami Council. They joined at different times, and activity rates vary considerably. For example, the Russian population groups have reduced their activity in recent years. On the other hand, the Sami Council has more resources and can therefore make a greater contribution to the work of the Council. The same applies to the North American organisations whose political legitimacy is far greater than that of the Russian groups. This among other came to show in Canada's work to ensure a central place for indigenous groups in the Arctic Council when it was established.

Which organisations should be admitted and how they would be represented were hot topics at the start of the Arctic Council.[33] Not surprisingly, the indigenous groups do not always agree among themselves either. Since the Nenets in Russia, the Sami in Scandinavia and the Inuit in Greenland face different challenges, it might be easier to implement smaller local projects than pan-Arctic ones, that is, where challenges facing the indigenous peoples in the north as a whole are addressed. They are not a unified group with coinciding interests and one voice, but they do have a community secretariat—The Indigenous Peoples' Secretariat (IPS), established in 1994. IPS has its own offices in Tromsø and a programme budget of its own too. Its mission is to support the work of all indigenous peoples represented on the Council, with, for instance, formulating project proposals. Funding of indigenous peoples' work in general and the IPS in particular, is, like the funding of working group projects, a constant challenge.

The fact that indigenous groups play such a significant role in international forums, such as the Arctic Council, is unusual. They are permanent participants and must be included in all debates on priority-setting, but they have no voting rights, unlike the member states. Nevertheless, this form of organisation highlights the very central, practical and symbolic importance of the groups in the alliance. And indeed, one of the formal criteria for obtaining observer status in the Arctic Council is support of the indigenous peoples of the Arctic.

* * *

The role of observers in the Arctic Council has been discussed since the initiation of the Arctic Environmental Protection Strategy (AEPS) in 1991 and the start of the Arctic Council in 1996. Several non-Arctic States, international organisations and NGOs have observer status in the Arctic Council. As of 2019, the number is 39, with representatives hailing from Switzerland to UN's environmental programme to WWF.[34] The role of observers is primarily to observe. Nevertheless, we must make a distinction between the SAO level and working group level. At the SAO level, observers have limited influence. We have seen from time to time that observers (and applicants) have excessive expectations of the role they can play in the forum. The SAOs have also taken steps to give observers a realistic picture of what they can do. Although observers can go through the member states or the permanent participants to propose new projects, their primary role is to offer ideas and opinions with regard to the projects already approved by the Council, that is, the projects of the working groups. This limited influence has occasionally resulted in frustration and may also have reduced the interest of observers in the work of the Council.

Although the role of observers has been debated since the birth of the Arctic Council, it was only during the Danish and Swedish chairmanship (2009–2013)—and especially in connection with the 2013 ministerial meeting in Kiruna—that the question finally made it to the top of the agenda. Increased interest in the Arctic in general and in observer status in the Arctic Council in particular led to an internal member state review of the rules that apply to observers, the criteria for accepting new observers and the procedures for their participation. The result was presented in an SAO report to the 2011 Nuuk Ministerial Conference and expanded in the revised Arctic Council Rules of Procedure[35] and the Arctic Council Observer Manual[36] adopted at the Kiruna Ministerial Conference in 2013. The purely formal requirements are also described in these documents; they also state the main role of the observers: to participate in the working groups.

To determine the eligibility of applicants, several criteria are considered. Three of them deserve particular attention. Firstly, observers must acknowledge and accept the legal framework that applies in the Arctic. This point is linked to the importance attributed by the Arctic Coastal States to the law of the sea as it applies in the Arctic. The law of the sea sets international rules on the ownership and management of marine resources. The economic zones are critical in giving Coastal States sovereign rights over natural resources within 200 nautical miles (370 km) from the coast. The Coastal State thus determines the manner in which, for example, fishery resources are to be managed and it has sovereign rights to potential petroleum resources. The law of the sea also stipulates how boundaries between the various areas of the sea should be drawn up and how the size and extent of the continental shelf must be determined. The law of the sea gives Coastal States a very special position.

The second point is that observers must be able to demonstrate relevant Arctic expertise. Here there is a certain interpretative leeway since "relevant Arctic expertise" is a vague term and can mean many different things. One last point is that observers shall work to enhance the efforts of permanent participants, that is, the indigenous groups. Again, the question of how performance in this regard can be measured remains unclear.

There are several arguments in favour of involving observers in the Arctic Council's work. First of all, involvement can increase awareness of the challenges facing the Arctic. It can make it easier for non-Arctic scientists to provide expertise. One last and important reason is that most of the Arctic challenges are global in character, such as climate change, environmental pollutants and shipping. These challenges can only be solved globally, not by the eight permanent members of the Council alone. The arguments against letting observers in on the work of the Council are both practical and political. As far as the practical reasons are concerned, meetings of ministers and government officials are usually held in remote places far off the beaten track. Simply getting to these places is time consuming and expensive. It can also be difficult to find conference rooms able to accommodate all the participants. This has been used as a reason for limiting the number of observers. In terms of politics, there are two main types of problem. First, the permanent participants are afraid that a large number of observers will marginalise the indigenous peoples' voice in the alliance. One could, however, argue that permanent participants would be supportive of a large number of observers if states increased financial support. Second, geopolitical turbulence and practical politics—*Realpolitik*—such as States' fear of losing power, have also affected the discussion about new observers. As already mentioned, several States, especially Canada and Russia, were sceptical about giving China access to the Arctic alliance, and due to EU sanctions against Russia, it was impossible to obtain Russia's agreement to grant permanent observer status to the EU at the 2015 Iqaluit ministerial meeting in Canada and the 2017 meeting at Fairbanks in Alaska. Again, we see how member states have the last word and that consensus is the bottom line.

* * *

While this book's length and scope make it impossible to undertake in-depth analyses of each member country, a short presentation is still necessary. It can tell us something about the Council's freedom of action. The Arctic States have different climatic, geographical, political and cultural conditions informing their policies on the north. All the Arctic States have moreover drawn up their own Arctic strategy papers. These documents have different political weight, and we need to be careful about leaping to over-zealous conclusions regarding their importance in terms of practical policies. Nevertheless, they can tell us something about general attitudes towards Arctic issues and, not least, the Arctic Council.

In 2017, Finland took over the chairmanship of the Arctic Council. Under their leadership, the expressed aspiration has been to work in areas such as education, meteorology and telecommunications, as well as core areas like the environment and climate of course.[37] This approach to Arctic issues is also evident in the Finnish strategy from 2013, where knowledge is the general theme. The emphasis on Finland as a nation facilitating education and promoting knowledge gives Finland an important role also in terms of the development of the Arctic.[38] Another point is the importance of the indigenous dimension, but also local populations as such. The creation of an Arctic identity is an important concern. Having closely studied official Finnish documents on the Arctic, we see a need on the part of Finland to tell the world that the country is an Arctic nation as well. Defining itself as part of the Arctic gives Finland the legitimacy to say how it thinks the region should evolve. Large parts of Finland are also located north of the polar circle, the classical definition of the Arctic region. Another point in the 2013 strategy deserving mention is Finland's clear articulation of how it intends to implement the different measures. Finland's policy on the Arctic and wider Arctic questions is also typical of the country. Its pride in its education and technological expertise shines through when it talks about the Arctic. An important dimension in Finnish policy on the Arctic Council is its unequivocal desire to have Arctic questions addressed under the aegis of the Arctic Council. The Finns are not interested in letting the Coastal States, often referred to as the Arctic Five or the A5, monopolise the Arctic debate and create forums from which other Arctic States (read Sweden, Finland and Iceland) are barred. Finland is just as "Arctic" as Norway in that, for example, all ports freeze over during the winter months. This is one problem Norway does not have.

Canada's Arctic policy is heavily informed by the indigenous dimension, along with the emphasis on allowing and persuading local communities to participate in economic development (politically and commercially). The human dimension is a constant element. Canada is huge geographically. But although 40% of Canada's territory is located north of the polar circle, only about 100,000 people live there, and 50% of them are indigenous. Canada's 2010 strategy displays a clear national bias and strong emphasis on sovereignty.[39] This does not mean that Canadians envisage different conflict scenarios, but national presence (also in terms of expanding the capacity of the armed forces) is emphasised relatively strongly. This was particularly evident under the Conservative government of Prime Minister Stephen Harper. Canada's rhetoric was sometimes rather acerbic; some might say it was confrontational with regard to the north. Security was and is the most important factor, followed by indigenous peoples. However, Canada has repeated on several occasions that the Arctic Council is the most important Arctic forum for cooperation, and it isn't difficult to perceive a sense of Canadian pride in the country's role in the creation of the Council. A good example is the document outlining the objectives of Canada's chairmanship from 2013 to 2015. The document starts by stating that Canada was the first State to occupy the chairmanship of the Arctic Council.[40]

However, when Canada took on the chairmanship of the Arctic Council in 2013, some feared the country had lost some of its initial enthusiasm. There was a parallel concern that the Council under Canada's chairmanship would be used mainly as a means of enacting particular policies at home.[41] They were right to worry, at least to a certain degree. Canada behaved as expected, and domestic policy objectives proved more important than global challenges such as climate change. However, when Prime Minister Stephen Harper lost the election to Justin Trudeau in 2015, the problems of climate change started to regain a more prominent place on the Canadian agenda. It would be too simple, however, to lay the blame for the reluctance to address pan-Arctic challenges, such as climate change and the environment, on the priorities of Harper's government during the Canadian chairmanship. At the time, international relations were strained and Russia's assertive policy in Ukraine affected both the international and Arctic agenda. And, as noted earlier, there was palpable frustration in the north too, which put a damper on the elaboration of wide-ranging pan-Arctic visions. The main point is that in recent years, Canada has maintained a rather local and national focus in its Arctic policy, including its behaviour in the Arctic Council.

Canada's neighbour to the south, however, is where most of the constraints on policy making in Canada originate. What the Americans think is important with regard to the Arctic, and Canadians are obliged to take that thinking into account. In official documents describing the ambitions of the US in the region, a great deal of space is devoted to protecting key national interests. Both the 2009 US Presidential Directive and the glossier strategy of 2013 begin by highlighting the country's wide-ranging and fundamental security interests.[42] This is evident in the foregrounding of missile defence, surveillance and strategic deterrence. A central and often repeated sentiment is the importance of the freedom of the high seas (military and commercial). The Northwest Passage, for example, is depicted as an international strait. Of the softer dimensions receiving particular attention in US Arctic policy, pollution (local and global) and oil spill preparedness can be mentioned. In this context, the importance of establishing transparent multilateral forums is highlighted. The positive attitude towards the 2011 search and rescue agreement and the 2013 oil spill preparedness agreement is a concrete expression of this mentality. During the 2015–2017 chairmanship of the US, the hard security dimension had to give way to a fairly concentrated focus on climate change and scientific cooperation in the region. At the same time, local measures were foregrounded such as, among other things, projects aimed at strengthening local communities in Alaska. There is limited space in this book to discuss US involvement in the Arctic. Nevertheless, it has been a political issue of no little importance, as testified by the attendance of US secretaries of state at the latest ministerial meetings of the Arctic Council.

The participation of Secretary of State Hillary Clinton at the 2011 Nuuk meeting and Secretary of State John Kerry at the 2013 Kiruna and 2015 Iqaluit meetings confirm the importance of Arctic issues on the US foreign policy agenda. Donald Trump's former Secretary of State Rex Tillerson took part moreover at the 2017 Fairbanks meeting. In addition to the growth in international interest in Arctic issues in terms of climate and socio-economic developments, there are several other reasons that explain the involvement of the US.[43] But US foreign policy engagement in the Arctic has fluctuated all the same. The US maintained a rather back-seat position before 2010, segueing to a more active period up to 2017, after which policy has remained fairly ambiguous.[44] Some point to the transition from a Republican to a Democratic administration in 2009 as the reason.[45] People with a long track record in the Arctic Council have also pointed to the personal involvement of the two secretaries of state of the Obama administration, for example, John Kerry's preoccupation with questions concerning the importance and health of the oceans coincides with a commitment to the Arctic.[46] The Macondo accident also lifted oil spill preparedness onto the political agenda in the US.[47] The country's commitment remained in place during the chairmanship of the US from 2015 to 2017.

Where the Arctic and climate-related questions fit in the new administration's list of priorities is, of course, hard to say, but it may be overoptimistic to expect the US to galvanise its involvement in the region. As the world's only superpower, and with a president whose attitude to international cooperation is relatively lukewarm, other challenges are likely to attract wider attention. The commitment of the US to the north must, however, always be seen in light of developments in the country on the other side of the Bering Strait.

Russia's stated Arctic policy is largely good news to anyone interested in a stable Arctic and multilateral cooperation on Arctic-related matters. The emphasis on Russian sovereignty is clearly accentuated, but it is not a surprising emphasis all the same when we know that Russia has 40% of the Arctic coastline. Like all other States with jurisdiction in the Arctic, they have a legitimate interest in the region. The law of the sea is a key factor and Russia's attitude to international cooperation is positive, as long as it doesn't threaten Russian sovereignty. An example is Russia's 2008 Arctic Strategy, where Russia states its intention in no uncertain terms to control the Northern Sea Route. In a historical perspective, however, the wording here is fairly laid back.[48] The Strategy sets out in plain language what Russia considers its legitimate need to exploit natural resources in the Arctic. Environmental issues are given some attention, but climate change in Russia's view is a natural phenomenon, not a man-made one. Denying the anthropogenic nature of climate change could create and has created complications in the work of the alliance. But at the same time we see a strong determination in Russia to adapt to a new Arctic. Russia has played a key role in the development of search and rescue coordination in the Arctic, as was evident in the efforts to negotiate the 2011 search and rescue agreement and the later 2015 oil spill agreement. Russia was one of the leading participants in the work resulting in the two agreements. It has therefore been easier to work with Russia on maritime safety than environmental and climate issues. At the same time, Russia is "hard power-oriented"—not surprisingly since the country rightly or wrongly believes it is surrounded by enemies in the north. In practice, however, Russia does work with others on Arctic-related matters. The country has also been active in drafting the research agreement. By and large, then, Russia has a positive attitude towards the Arctic Council. As in the US, the Arctic is perceived as an area where conflict is unlikely, and where there is a common interest in keeping it that way.

Norway has been one of the most forward-looking States with regard to the "new" Arctic, and its stated commitment to sustainable development in the north is perhaps the most comprehensive national political project of recent years. Where Norway can be said to differ from the other member states is in the breadth of its engagement. Norway has interests in virtually every field of significance to the development of the Arctic, including fisheries, petroleum and shipping. The Gulf Stream makes Norway unusual and somehow a little less Arctic than the others. Less ice and a warmer climate allow for a more active use of Arctic areas.

Although Norway was not particularly active in the first years of the Council, the Council has attracted more and more political interest, especially in recent years. With the establishment of the permanent secretariat in Tromsø and the location of three of the working group secretariats (AMAP, ACAP and EPPR) in the same city, Norway seems to have assumed a sort of leader role.[49] Some of the other member states have responded to this development with a certain amount of scepticism, but Norway has been able to pledge resources and has worked hard diplomatically to consolidate its position.[50] Norway has also led several major projects with others under the auspices of the Arctic Council, especially projects in which AMAP has had lead responsibility. The fact that the AMAP secretariat has been located in Oslo (from 2018 in Tromsø) and that the activities of this working group are largely funded by Norway does help explain the situation. Furthermore, Norway has lobbied to have China and the EU awarded permanent observer status.[51] In the run up to the meeting in Kiruna in 2013, Norway also expressed support for the EU's bid to become a permanent observer. A recurring argument has been that parallel and competing institutions should not be permitted in the region, something that might well come to pass if no other stakeholders with a commitment to the Arctic are allowed to apply for observer status. An important point is that the Norwegian authorities want the Arctic States (and especially Coastal States) to be in the driver's seat. The Arctic Council supports, as previously mentioned, the primacy of the United Nations Convention on Law of the Sea (UNCLOS) as it applies to the Arctic Ocean.

Whether this effort on the part of the Arctic Council is rooted in a clear and well thought out strategy is somewhat unclear. But what we have witnessed is an active policy to make Tromsø the Arctic "capital" and to strengthen international cooperation in the north.[52] Norwegian authorities refer to the Arctic Council as the leading political body in a region in which Norway has significant strategic and economic interests. It is also important to bear in mind that Norway is a northern offshore oil producer and thus a stable working environment is important. At the same time, Norwegian authorities have worked hard to ensure as high a place as possible on the agenda of the Arctic Council for multilateral collaboration on environmental matters. This can be seen in the use by the environmental administration in Norway of AMAP's work in international negotiations.[53] This environmental profile is also evident in the amount of money Norway invests in joint Arctic projects. Most of it goes to multiparty environmental work in general and to AMAP in particular. A few changes can be seen here, however, as Norway spends just as much on various secretarial functions, especially the secretariat in Tromsø, where Norway is the State that transfers most resources.

The approach of the smaller States with jurisdiction in Arctic waters, that is, Iceland and Denmark, is one which underlines the importance of a well-functioning governance regime which also confers on them sovereign rights. For example, the Danish 2011 Arctic Strategy devotes a great deal of space to the law of the sea.[54] This is particularly perceptible in its comprehensive description of the work to document the outer limits of the continental shelf. With regard to Coastal States' rights, a more concentrated alliance of Arctic States (A5) is mentioned. In the 2016 Danish Foreign Policy Annual Report, Denmark's main political objective is described as strengthening "Denmark's position as an Arctic player and pursuing the goal of maintaining the Arctic as a low-tension area through enhanced international cooperation with a particular focus on marine cooperation, telecommunications infrastructure and research partnerships".[55] There was a debate in Denmark in 2016 about the possibility of the Arctic Council getting to address security issues. As mentioned earlier, the idea was not welcomed by the political leadership. An interesting observation in this context is that it was the Danish authorities who initiated the 2008 meeting of the five Arctic Coastal States in Ilulissat, a forum States without a coastline north of the polar circle, demonstrated little enthusiasm for. The meeting format has not really been formalised even though the Arctic Coastal States came together again in 2010 and in the spring of 2018. The other Arctic States were invited to the latter meeting, which also served to mark the tenth anniversary of the Ilulissat era.[56] Again, the most important thing was to emphasise the agreement of the member states on the basic governance mechanisms in the north (read: Convention on the Law of the Sea) and that the region should remain a low-tension area.

While relatively little space is devoted to traditional security challenges, maritime safety is, conversely, given considerable attention. The main theme in Danish Arctic politics is nevertheless the development of Greenland (including the Faeroes, albeit to a lesser extent). Key words are mineral extraction and petroleum activity. If we read the 2011 Strategy, which extends until 2020, it is the "new" industries in the future Arctic that catch absolutely most of the attention of the authors. What is noticeable here is the lack of attention to fishing and hunting, traditionally the most important industries in Greenland. Economic development in the Arctic region has been a priority, which tallies with Denmark's approach as expressed in the Arctic Council. A key point—which should not really come as a surprise—is that relations with Greenland define Copenhagen's approach to Arctic policy: without Greenland Denmark is not an Arctic State.

Iceland is a small State, a fact that affects its approach to the Arctic. Iceland's Strategy or rather the 2011 Parliamentary Resolution is the shortest and least elaborate of all the Arctic "Strategies".[57] A relatively large amount of space is devoted to the importance of the Arctic Council as a collaborative arena. The Strategy advises against allowing the A5 alliance to achieve a more prominent role, cf. the 2008 meeting at Ilulissat. Politics as a whole in Iceland is nevertheless influenced by the demanding economic situation the country found itself in after the economic crisis of 2008. For example, less was spent on international cooperation (although the PAME and CAFF secretariats, both located in Iceland, are still largely funded nationally). This was evident, for example, in Iceland's customary opposition to increases in the joint budget of the Arctic Council secretariat, and its expressed scepticism about the relocation of the IPS (Indigenous Peoples' Secretariat) from Copenhagen to Tromsø. These attitudes are changing, however, and Iceland is again an active Arctic player. A final point of interest, one which also substantiates greater activity on Iceland's part, is the country's portrayal of itself as a venue for international meetings and conferences on Arctic issues. Here, Iceland both can and should play a more central role, as reflected in the commitment to the Arctic Circle International Conference, which some have mistakenly perceived as a competitor to the Arctic Council.[58] Iceland has in this regard found its own special niche. We can see some of the same trends in Sweden.

Sweden marked its accession to the chairmanship of the Arctic Council in 2011 by publishing its Arctic Strategy.[59] The importance of the Council as the Central Multilateral Arctic forum is pressed home repeatedly. As in the Icelandic Strategy, Sweden is unhappy about any expansion of the A5 group and formalisation of the alliance. Sweden has no petroleum interests in the region as such, although it can supply subcontractor services to the industry. More important, and launched as one of three main priorities of Swedish Arctic policy, is the need for action to address the climate and environment. Important challenges requiring action in collaboration with other Arctic States include limiting the emission of short-lived climate gases and environmental pollutants. As in the strategies of some of the other Arctic States, the indigenous population dimension receives relatively large coverage. This is reflected not least in the fact that the human dimension is the third main priority of the 2011 Strategy. Little has changed here. Energy and minerals in the Arctic are not a big issue in Sweden, and the pressure on dealing with climate and environmental issues will continue to be felt in the work of the Council. It is also interesting to observe Sweden's long-standing support for the establishment of protected areas in the Arctic. Here, what exactly constitutes common interests in the Arctic is not obvious, and in countries with a coastline and focus on national and local economic development, people are not always receptive to these matters. The point is, however, that Sweden (and Finland too) has been, for self-evident reasons, more interested in the environment and climate in the north. Neither Sweden nor Finland are, after all, Arctic Coastal States, with the rights and obligations that entails.

We need to be careful before drawing far-reaching conclusions on the basis of this brief presentation of key elements of the Arctic States' attitudes towards the Arctic and collaboration in the Council. Nevertheless, we can briefly summarise by saying that the big States (geographically and militarily) are concerned with security. It is hard, moreover, to be "Arctic" without mentioning the subject of indigenous peoples. One last observation: States without sovereign rights in Arctic waters are more positive to the creation of protected areas—a not particularly surprising discovery since they have less to lose. I return to these contrasting views in the next chapter. But the overall impression is that attitudes towards the Arctic Council's role as a political and scientific venue for discussions on Arctic challenges and opportunities have been positive in recent years. The Arctic Council is considered by each of the Arctic States as the most relevant forum to discuss Arctic issues, as can be seen in their respective strategies for the region.

* * *

One of the main insights we have gained is that the Arctic Council has created an arena where different players can come together and networks are created. The permanent participants' position and role are critical. They have been involved in a number of projects, ACIA being the best known. At the same time, scientists have helped shape the Arctic agenda even their influence does vary from field to field depending on how politically controversial proposals have been and, of course, how expensive. They have nevertheless had to be heard and have been able to provide decision-makers with the relevant scientific basis on which to take political action. The observers have also been members of this network. While not everything runs smoothly, and much is still unclear in relation to how and how far observers may take part, we find several examples of observers playing a role and helping to strengthen this network of actors that defines the Arctic Council. One of the most prominent observers is WWF. They have "used" the Arctic Council in a variety of ways. The literature on the influence of NGOs often distinguishes between different roles. NGOs can set the agenda, provide information, formulate draft agreements and so on. In the Arctic Council, WWF has performed several of these roles. An interesting example is in the field of oil spill preparedness. In the work on the oil spill preparedness agreement, WWF went to the media directly because, in their view, the deal had little to say about the real problems, including heavy oil. Despite the fact that WWF had only moderate influence on the agreement, they were included in the process as members of the task force that drafted the agreement and were therefore on the inside. This gave them legitimacy, at the same time as WWF's inclusion expanded the agenda. A similar example is the work done on preventative measures in the field of oil spills. Taking Norway, the standards for preventative action are determined by the industrial sector, and with regard to shipping, regulations are formulated mainly at the national level and at the UN International Maritime Organization (IMO). In this case, NGOs (primarily WWF) felt they had to go outside Arctic Council if they were to make a difference. The Arctic Council's work in this field has thus been largely confined to compiling information, and there have been few possibilities to take a proactive approach by, for example, negotiating an agreement on oil spill prevention in the north. Some of the interesting stakeholders to emerge in this connection are the various businesses with interests in the north. In historical terms, they have never been important members of the network created by the Arctic Council, nor has the industry been particularly involved in the drafting of preventative measures. However, it is not entirely wrong to say that on the Norwegian and Russian side (the only two States with petroleum fields offshore), the State has spoken on behalf of the industry. Their indirect influence is therefore plain to see. So while commercial actors have been more or less absent from the work of the Arctic Council, they have influenced the policy choices of States in discussions about the future of the Arctic. The establishment of the Arctic Economic Council (AEC) is evidence of this trend. It was necessary to create an arena outside the Arctic Council to encourage the wider involvement of commercial actors. All the same, the idea of the Arctic Council as a network is largely accurate. This does not mean that players with less influence (read observers, permanent participants and scientists) decide the general direction of the alliance, but the member states cannot ignore them. It is at this tense intersection where member states' desire for political control confronts a wide range of actors with different agendas and different opportunities to influence issues of relevance to Arctic development, that the Arctic Council is established.

Notes

1. Arctic Council (1998) "Arctic Council Rules of Procedure". Available from https://oaarchive.arctic-council.org/handle/11374/940. Last visited 25 April 2019. The procedures were last updated in 2013.
2. The procedural rules can thus be changed by consensus. The Arctic Council can be different from what it is today if that is what the member states want. There is a formal dynamic here. Ibid. Art. 48.
3. Previously, the secretariat rotated between the member states.
4. That said, the SAO meetings often have a rather broad agenda. There are many things to discuss. The meetings have thus been criticised for being slightly too formalistic and of doing little to foster good, practical discussions.
5. An exception is the American SAO, Julie L. Gourley, who has been in post for over ten years.
6. Heather Conley and Matthew Melino (2016), "An Arctic Redesign Recommendations to Rejuvenate the Arctic Council", CSIS Report. p. 6.
7. Olav Schram Stokke: "An inner circle in the Arctic Council?" Opinion piece in *Nordlys* 28 April 2013.
8. Two of the most active researchers are Timo Koivurova and Oran Young. See, for example, Timo Koivurova (2010): "Limits and possibilities of the Arctic Council in a rapidly changing scene of Arctic governance", *Polar Record*, 46 (2), pp. 146–156, Oran Young (2009) "Whither the Arctic? Conflict or cooperation in the circumpolar north". *Polar Record*, 45 (232), pp. 73–82. and Oran Young (2011) "If an Arctic Ocean treaty is not the solution, what is the alternative?" *Polar Record*, 47 (4), pp. 327–334.
9. Paula Kankaanpää and Oran Young (2012): "The effectiveness of the Arctic Council". *Polar Research*, 31.
10. The biggest conference organised by the Arctic Council was, on the other hand, under the auspices of CAFF and took place in Trondheim in the autumn of 2016.
11. For an overview of this work, see: Christian Prip (2016): "The Arctic Council and Biodiversity Need for a Stronger Management Framework?" *Nordisk Miljörättslig Tidskrift*, No. 2, 2016, pp. 37–53.
12. Ecosystem services is a collective term for the basic goods and services we can harvest from nature, such as food and other resources, but also outdoor activities and the like.

13. Christian Prip (2016): "The Arctic Council and biodiversity need for a stronger management framework?"
14. The debate on protected areas is also linked to the discussion of a possible "Arctic treaty", inspired in part by the Antarctic Treaty. The Arctic Coastal States refuse to contemplate any such idea (see Chap. 5).
15. Christian Prip (2016): "The Arctic Council and biodiversity—need for a stronger management framework?".
16. IMO is the UN's maritime safety organisation and was established in 1948 to promote maritime safety and prevent pollution of the marine environment. IMO has 170 member states and a number of specialised committees drafting international legislation and regulations. For more information, see: http://www.imo.org/en/Pages/Default.aspx.
17. IMO adopted an international code for ships operating in polar regions in November 2014. The Polar Code provides binding rules for the protection of ships, crews and passengers in challenging polar waters. For more information, see: http://www.imo.org/en/MediaCentre/HotTopics/polar/Pages/default.aspx.
18. This work is highlighted on EPPR's website, see: https://www.eppr.org/sar/ and http://www.eppr.org/mer/.
19. In 2018 an external evaluation was conducted. It is, however, not available for the public.
20. SDWG (2004), The Arctic Human Development Report, Available at https://oaarchive.arctic-council.org/handle/11374/51?show=full. Last visited 25 April 2019. SDWG (2015) The Arctic Human Development Report II, Available at https://oaarchive.arctic-council.org/handle/11374/1653?show=full. Last visited 2 April 2019.
21. Lassi Heininen, Heather Exner-Pirot and Joël Plouffe (2016), The Arctic Council: 20 Years of Regional Cooperation and Policy-Shaping, *Arctic Yearbook*, p. 11.
22. For a more comprehensive list, see: https://www.sdwg.org/activities/project-reports-from-completed-sdwg-projects-1998-to-2015%20/.
23. Ida Folkestad Soltvedt and Svein Vigeland Rottem (2017): Challenges of the Arctic Council's Sustainable Development Working Group: How to Improve? *FNI Report* 2. Lysaker, FNI.
24. Pekka Haavisto (2001), Review of the Arctic Council Structures. Available at https://oaarchive.arctic-council.org/bitstream/handle/11374/449/ACSAO-FI01_6_AC_Structure_final.pdf?sequence=1. Last visited 25 April 2019. and the Office of the Auditor General's *Undersøkelse av myndighetenes arbeid med Arktisk råd, Dokument 3:3* (2014–2015).
25. Ibid.

26. For a complete list, see the Arctic Council website: https://arctic-council.org/index.php/en/about-us/subsidiary-bodies/task-forces. Last visited 25 April 2019.
27. Svein Vigeland Rottem (2016): The Arctic Council: Vision, Structure and Participation. *FNI report* 4. Lysaker, FNI.
28. Two expert groups with a more limited mandate have been appointed. They are the Expert Group in Support of the Implementation of the Framework for Action on Black Carbon and Methane and the Ecosystem-Based Management Expert Group (terminated in 2013). The establishment of these groups follows the same pattern as the establishment of task forces, and therefore will not be dealt with separately here. It can also be mentioned that the working groups operate with expert groups with a limited mandate. There isn't the space in this relatively short description of the Arctic Council to pursue these groups further.
29. Svein Vigeland Rottem (2016): The Arctic Council: Vision, Structure and Participation. FNI report 4. Lysaker, FNI.
30. Ibid.
31. Arctic Council (1998). "Arctic Council Rules of Procedure".
32. It is possible that others can be included. Ibid.
33. For a detailed account of this discussion, see English: *Ice and Water. Politics, Peoples, and The Arctic Council.*
34. For a complete list, see: https://arctic-council.org/index.php/en/about-us/arctic-council/observers. Last visited 24 April 2019.
35. Arctic Council 2013, The Revised Arctic Council Rules of Procedure. Available from https://oaarchive.arctic-council.org/handle/11374/940. Last visited 25 April 2019.
36. Arctic Council 2013, The Arctic Council Observer Manual (last updated 2016). Available from https://oaarchive.arctic-council.org/handle/11374/939. Last visited 25 April 2019.
37. Arctic Council (2017), Exploring Common Solutions, Finland's Chairmanship Programme for the Arctic Council 2017–2019.
38. Finland (2013), Finland's Arctic Strategy. Available from https://um.fi/finland-s-arctic-strategy-and-northern-policy. Last visited 25 April 2019.
39. Canada (2010), Canada's Northern Strategy. Available from https://www.northernstrategy.gc.ca/index-eng.asp. Last visited 25 April 2019.
40. Canada (2013), The Canadian Arctic Council Chairmanship. Available from https://arctic-council.org/index.php/en/our-work2/8-news-and-events/226-canadian-chairmanship-program-2013-2015. Last visited 25 April 2019.
41. These worries were expressed in a series of interviews I conducted in connection with a project on Norway's role in the Arctic Council in the winter and spring of 2013. In the view of the Norwegian Ministry of Foreign

Affairs, it was an issue of prime importance in its work with the Council. Interview at the Ministry of Foreign Affairs 30 April 2013.

42. US (2009) HSPD 25: Arctic Region Policy. Available from https://www.hsdl.org/?abstract&did=232474. Last visited 25 April 2019 and US (2013) National Strategy for the Arctic region. Available from http://polarconnection.org/us-national-strategy-arctic-region-may-2013/. Last visited 25 April 2019.
43. For a detailed account of US involvement in the Arctic Council during this period see in addition Torbjørn Pedersen (2012): "Debates on the role of the Arctic Council".
44. Nevertheless, the US, it should be noted, is the member state that has attended most meetings under the auspices of the Arctic Council. For a breakdown of the participation of member states, permanent participants and observers in the Arctic Council, see: Sebastian Knecht (2016): "The Politics of Arctic International Cooperation: Introducing a Data Set on stakeholder participation in Arctic Council meetings, 1998–2015" in *Cooperation and Conflict*, 52 (2), pp. 203–223.
45. Interview with a member of the EPPR team 22 April 2013 and AMAP team 02 May 2013.
46. Interview with Ministry of Foreign Affairs, 30 April 2013.
47. Interview with Ministry of Fisheries and Coastal Affairs, 27 June 2013.
48. Russia (2008), Russia's Arctic Strategy: Ambitions and Constraints. Available from https://archive.org/stream/563663-russias-arctic-strategy/563663-russias-arctic-strategy_djvu.txt. Last visited 25 April 2019.
49. Mention can also be made of Norway's eagerness to get the newly created Arctic Economic Forum (AEC) and the Indigenous Peoples' Secretariat (IPS) located in Tromsø. Norway, of course, is an important economic contributor.
50. There is no complete breakdown of who finances the working groups or how it is done, but several of the people I have interviewed in recent years describe Norway as the most important financial contributor to the Arctic Council and the working groups. This commitment was evident during the important formative period around 2013, when interest in the Arctic among non-Arctic States was particularly visible. Interview with Ministry of Foreign Affairs 30 April 2013, interview with AMAP 02 May 2013, interview with EPPR 22 April 2013. See also Elana Wilson Rowe (2013): "Arctic hierarchies? Norway, status and the high north". *Polar Record*. Published online February 2013. This trend seems set to continue, with Norway being the main funder of the permanent secretariat in Tromsø.
51. This despite the diplomatically tense relationship between China and Norway after the Peace Prize was awarded to Liu Xiaobo in 2011. Norway

wanted, for example, China to be given status as a permanent observer in 2011. Interview with the Ministry of Foreign Affairs 30 April 2013.

52. The permanent secretariat was formally opened during the Arctic Frontiers conference in the winter of 2013. During the joint Nordic presidency from 2006 to 2013, there was a temporary secretariat in Tromsø. The opening of the permanent secretariat itself took place in May 2013.
53. Svein Vigeland Rottem (2017): "The Use of Arctic Science: POPs, Norway and the Stockholm Convention", *Arctic Review on Law and Politics* 8.
54. Svein Vigeland Rottem (2017): "The Use of Arctic Science: POPs, Norway and the Stockholm Convention", *Arctic Review on Law and Politics*, 8.
55. Danish Ministry of Foreign Affairs, Annual Report 2016. Available from http://um.dk/en/. Last visited 25 April 2019.
56. For a more detailed analysis of this initiative, see Jon Rahbek-Clemmensen and Gry Thomasen (2018): "Learning from the Ilulissat Initiative: State Power, Institutional Legitimacy, and Governance in the Arctic Ocean 2007–2018", University of Copenhagen, Centre for Military Studies.
57. Iceland (2011), Arctic Policy of Iceland. Available from https://arcticiceland.is/en/stefna-islands-i-malefnum-nordhurslodha. Last visited 25 April 2019.
58. However, where competition is in fact greater is between Arctic Frontiers and Arctic Circle. It is no secret that Norway has wanted to come across as the optimal place to be in when the future of the Arctic is discussed. The idea of Tromsø as the Arctic capital has led to the location of the Arctic Council's secretariat here, together with secretariats of AMAP, ACAP and EPPR, as well the Indigenous Peoples' Secretariat (IPS). Nothing should interfere with this thought. The Arctic Circle is thus rightly or wrongly perceived as a competitor.
59. Sweden (2011), Sweden's Strategy for the Arctic Region. Available from https://www.government.se/country-and-regional-strategies/2011/10/swedens-strategy-for-the-arctic-region/. Last visited 25 April 2019.

CHAPTER 3

The Arctic Council in Arctic Governance

Abstract In this chapter the Council's role in the governance of the Arctic is discussed. The working groups have accumulated unique expertise through their regular assessment of Arctic issues. However, the agreements and guidelines created under the umbrella of the Council are limited in scope. In this chapter I argue that the Council will remain to all intents and purposes a decision-shaping body rather than a decision-making one. The Arctic Council is, however, a convenient and appropriate venue where several aspects of Arctic policy can be drafted in close consultation with key stakeholders. Of equal importance is its position as a producer of knowledge within the wider patchwork of international bodies whose work affects the Arctic.

Keywords Arctic Council • Arctic governance • Working groups • Knowledge

The first two chapters provided an overview of what terrain the Arctic Council is moving in, the organisation of the Council and the stances of the member states on developments in the Arctic in general and the Arctic Council in particular. The question in this chapter is from a political and legal perspective what type of forum we are dealing with and what scope of action the Council has. In a power-political analysis of opportunities and limitations, the starting point is that of the eight permanent members, five are allied in NATO, two are neutral[1] and Russia is the "balancing

S. V. Rottem, *The Arctic Council*,
https://doi.org/10.1007/978-981-13-9290-0_3

power". This is of course important for being able to understand the possibilities of the Arctic cooperation. In terms of a real-political logic, states will always be concerned with the distribution of power and will be suspicious of the motives of others. They will be obsessed with not losing power and hence their position in international politics.[2] States will thus be sceptical to cooperate if they fear that others will benefit more from the cooperation, and that it may possibly restrict the scope of the state to take action, or require them to implement measures that are perceived as politically and economically costly. The question is thus which topics may be subject to discussion and ultimately setting standards for how the individual state can "behave" in the Arctic. In general, we can say that the Arctic Council has kept itself to dealing with harmless topics, where there has been agreement on a given problem and how it should be dealt with. For example, everyone can agree that pollution in the region is not good, and when it is cross-border it can make sense to find common solutions.

At the beginning of the 1990s, as previously mentioned there was a discussion about whether the Arctic Council should also address issues of a security-policy nature. Finland and Canada were among the advocates for wider cooperation. A topic that has also been discussed later on, including in 2016 when the American think-tank Center for Strategic and International Studies (CSIS) published a report in which the authors argue that the Council had not to a sufficient extent dealt with security policy questions. They argue that the Council should be more involved in this regard.[3] Finland had funded this work, without officially endorsing the recommendations. The report received a lukewarm reception when it was presented in Washington in April 2016. In the spring of 2016, this was also discussed in Denmark in connection with the Danish diplomat Peter Taksøe-Jensen presenting a report to the Danish Prime Minister, Foreign Minister and Defence Minister. He aired similar ideas.[4] Again, the political leadership was rather restrained. One last example is from Canada where former foreign minister Lloyd Axworthy believed that this was the way to go, but once more this was a dead ball.[5] This tells us that time is not ripe for classic security policy questions to be addressed in this arena. It probably makes sense. The basic thinking of those who have promoted this proposal has been that by establishing a meeting place to discuss security policy questions, it can help curb a potentially increasing level of conflict in the region. The arguments for such an expansion of the Council's portfolio have also been linked to the fact that a large proportion of the challenges and especially the environmental challenges in the Arctic are closely linked to the military capacities of the Arctic states, both as potential sources of pollution, but also as tools for cleaning up (e.g. the Coast Guard). On the other hand, several have argued for the benefits of having an arena for collaboration on common challenges that are not linked to security policies. It is precisely the fact that the Arctic Council does not deal with security policy issues that has meant that the cooperation has been continued, including in turbulent times. An obvious example is that all security policy cooperation between NATO and Russia was frozen after the annexation of the Crimea. If the Arctic Council had security policy on its agenda, the same would have happened here. It is far from certain that it would have been possible to distinguish between the monitoring of environmental pollutants and conversations about military capacities. The point is that a meeting place has been established where among others, the Americans and Russians have the opportunity to meet during geopolitically strained periods. With this as a starting point, the Arctic Council cannot be an arena for security policy. If it had then it would have been something different, and maybe not even existed. The understanding is that the Council does not operate outside the logic of international policy, but can act as an alternative arena.

* * *

Another central observation is that five of the member states are coastal states. This means that due to the United Nations Convention on the Law of the Sea, these states have a completely different set of rights and obligations in terms of resource utilisation and management in Arctic sea areas than non-coastal states. It is precisely this difference between the states that have coastal state rights, often known as the Arctic Five (A5), and all the eight Arctic states together (A8) that is a recurring topic of discussion in debates about the Council's sphere of possibilities. The three states (Iceland, Finland and Sweden) that do not have coastal state rights in the Arctic Ocean have, as previously shown, expressed concern that Arctic questions have been dealt with outside the Arctic Council structure. The most well-known example is the meeting in Ilulissat, Greenland, in 2008, where the five coastal states met to discuss issues of particular relevance to them, such as the status of the Law of the Sea in the Arctic.[6] Following the meeting, they made a united declaration: "We remain committed to this legal framework and to the orderly settlement of any possible overlapping claims." After this meeting, the atmosphere was lacklustre among the Arctic states that did not participate, which could also be clearly seen in Iceland's Arctic "strategy" from 2011. It states that if consultations between the A5 develop further, this may lead to the Arctic Council becoming severely weakened.[7] In Finland and Sweden, attention is drawn to the Arctic Council as being the central forum for the consideration of all Arctic questions, and we need to be wary about institutionalising a cooperation in which only the Arctic coastal states sit around the table. Finland and Sweden are also more positively tuned to the idea of working together to establish larger protected areas in the Arctic. Not surprisingly, the coastal states that have a commercial interest in exploiting the resources that may exist in the sea areas off their coasts are more restrained. It is and will be a contrast that will characterise further debate on the future of the Arctic.

What we can nevertheless see is that the Arctic Five are in need of additional meeting places, where they can, for example, discuss the extent of Arctic continental shelves, coastal infrastructure and simply underline that they all agree that the Law of the Sea also applies to the Arctic Ocean. Professor and Arctic expert Olav Schram Stokke has said this in the following way: "The Arctic Five have in short a lot to talk about, and it is valuable that they create suitable spaces rooms for themselves to meet."[8] But even though there may be good practical reasons why the five meet, it does not mean that it is necessarily politically wise to establish parallel structures. This may be part of the reason why this format has not been further formalised, and meetings have instead been more sporadic. In 2018, for example, the foreign ministers met to repeat the undertakings and to celebrate the tenth anniversary of the declaration from 2008. They have also chosen to meet when case-specific topics have required it, such as in questions about fisheries in the Arctic Ocean (see below). It tells us something about the room to manoeuvre that the Arctic Council has. The Arctic Council may not only be a commonality of interest for the Arctic coastal States, but must constantly maintain the equilibrium between the A5 and A8. This has consequences for what the agendas for the Arctic Council's meetings will look like.

* * *

A third point is that the Arctic policy agenda differs from state to state, which is in addition to the relationship between the A5 and A8. In Canada, the indigenous population dimension has stood strong, which it still does. In Norway with completely different climatic conditions than the other

Arctic states, the issue of general economic development of the region is further ahead. The bottom line is that there is no one Arctic. An ice-covered Canada, a Greenland with a lack of infrastructure and relatively warm Northern Norway all pose a set of completely different challenges when developments in the Arctic are discussed. Internal economic matters, of course, also play a central role. For example, the fact that Iceland in the aftermath of the 2008 financial crisis had to put the brakes on is not surprising. It is probably also a widespread belief that Norway can afford to pay a little more. The States have different ambitions and wishes for the cooperation, which is particularly evident in the chairmanship periods. For example, Canada had an extensive focus on indigenous issues. Norway believed that this overshadowed challenges of a pan-Arctic nature, such as climate and the environment. Russia has been active in terms of maritime safety (search and rescue and oil pollution), but more reticent in terms of climate and especially emission limitations. Within the framework of this book, it will be impossible to outline all the conflicts of interest within various areas of cooperation in the Arctic Council. The overall acknowledgement is that the states have different goals and interests, which leads to the development of more and more projects, and thus also a sprawling project portfolio. This also leads to all the states wanting to focus on their pet projects during their chairmanship periods. This has two consequences. Firstly, there will be more and more projects in the Arctic Council to coordinate and follow up. Projects that often overlap. Secondly, there is a risk of losing continuity because projects are forgotten when a given state's chairmanship period is over, and that a lot of resources are used on something that is not followed up. The ambitions of the states form the framework for what the Arctic Council can do. The Arctic Council cannot be anything other than what the states make it.

* * *

The overall description should form the basis for any analysis of the Arctic Council. By extension, one should also look at what kind of legal status the Arctic Council has. The Council is a consensus body, where all the states must be in agreement before action is taken and projects are implemented. It is not an international organisation that can make binding decisions. For example, the three agreements negotiated in recent years are not formal agreements under the Arctic Council, but agreements between the eight member states. An international organisation is often bound by a treaty, it usually has a separate body that can operate more or less independently of the organisation's members and it is established and recognised within international law. The Arctic Council has none of these characteristics. The Council cannot operate independently of the states that established it, and it cannot sanction states that do not follow the guidelines formulated by the Council. Nor does it have a separate body that can follow up on states that do not implement measures that have been agreed upon. The three agreements that have been negotiated within the framework of the Arctic Council should, nevertheless, indicate that the Council is headed towards becoming an international organisation that also negotiates legally binding agreements. It is still a somewhat simplified analysis. We can look at the first two agreements signed in 2011 and 2013 respectively. How precise are they, are there mechanisms to follow them up, who helped to negotiate them and is it politically and economically costly to follow them up?[9] It is somewhat early to say anything about the research agreement from 2017, since at the time of writing it has only recently been signed. A closer presentation of the first two can also say something about which political areas can be negotiated within the framework of the Council and which fall outside. The agreement on search and rescue is an agreement with a defined objective. It is an agreement about the exchange of information and exercises. The objective of the agreement is "to strengthen aeronautical and maritime search and rescue cooperation and coordination in the Arctic".[10] The agreements describe the geographical responsibilities of the individual states in search and rescue.[11] We also find a list of relevant national points of contact. There are three matters that deserve particular attention here. Firstly, the agreement is similar to previous international agreements on search and rescue.[12] The agreement therefore confirms commitments to a greater extent than it gives new ones. Secondly, and unlike many other international agreements, new institutions are not established. The member states shall meet to discuss practical cooperation (article 10), but beyond that it is not very specific. A third point is that article 9 emphasises the importance of training and exercises, but as is made clear from Article 12, it is the states that will pay. There is therefore no financial strengthening of the work on search and rescue. It is thus a rather imprecise agreement with very little about any monitoring mechanisms. Despite having established expert groups in the EPPR in recent years to follow up the work on the agreement, the wording of the agreement has not changed. There is also no financial obligation in the agreement, and it was therefore probably quite easy to sign. The question is whether it could be wider, that is, include something about capacities and presence. Are the eight member states in the Arctic Council willing to go in that direction? There is little evidence of this. Increasing capacity is a national issue, and the Arctic Council cannot and will not make provisions in this respect.

Such an agreement could probably just as well have been negotiated between the Arctic coastal states, but by doing it within the framework of the Arctic Council, it was not necessary to start a larger process on the margins of the Council. The expertise that has been gained in the working groups made the Arctic Council a natural arena for such an agreement. This does not automatically mean that it is therefore very ambitious.

The oil spill agreement is fashioned in the same way as the search and rescue agreement. Article 1 refers to the purpose of the agreement: "to strengthen cooperation, coordination and mutual assistance on oil pollution preparedness and response in the Arctic".[13] This agreement is, however, more comprehensive than the search and rescue agreement. This can be seen in a non-binding appendix that gives instructions on how to coordinate and cooperate on any oil pollution in the Arctic. However, as with the search and rescue agreement, no additional resources are allocated to this work, and it is part of a wider group of international agreements.[14] This has led to criticism from, for example, WWF that the agreement does not say anything about which technology and what resources the states should have. One final point is that even here, no new formal structures have been established. It is the member states that will follow up. The agreement can thus be criticised for being rather imprecise, and does not add up to any financial obligations. Again, an agreement that was fairly easy to sign.

Both agreements fit well within the framework of the Arctic Council. For several years, considerable work has been done under the auspices of the Arctic Council, which is relevant to the agreements. By addressing the necessary environmental measures in light of increased commercial activity in Arctic waters, the working groups have acquired unique competence. Again, however, these are essentially agreements on information exchange and exercises. The agreements are limited in their range. It is therefore very strong to say that these are signs that the Arctic Council is on the way to becoming an international organisation with instruments that allow it to follow up on the work of the states in the respective fields. One of the main reasons why the agreement does not establish guidelines for petroleum activities in the region is that the five coastal states do not want to tie up their sovereignty in a sector of strategic importance.

Moreover, it can be argued that the regulation of the operations of commercial actors (petroleum, shipping and tourism) should take place at international level. It is not only the members of the Arctic Council who are active in the north. A global and not a regional set of rules is therefore a necessity.

* * *

So what do the two agreements from 2011 and 2013 tell us about the prospects for the Arctic Council continuing as a platform for negotiating legally binding agreements of relevance to developments in the Arctic and ultimately as an international organisation with stronger sanction opportunities?[15] There are three fields in particular that have gained attention in recent years: fisheries, biodiversity/protected areas and emission reduction measures, and then in particular of so-called short-lived greenhouse gases. In this context, one can argue that the Arctic Council can both be a too narrow and too broad platform for deeper cooperation, that is, for the development of legally binding agreements.

If we look at fisheries, the management of these resources is based on the migration patterns of fish stocks across the states' economic zones. Regional and often bilateral (such as the Norwegian-Russian Fisheries Commission) will therefore be more natural arenas for cooperation. If the international fishing fleet were to move further into the Arctic Ocean, the establishment of new regulatory mechanisms may be necessary. Such work has just begun. On 16 July 2015, Norway, Russia, the US, Canada and Denmark/Greenland signed a declaration against unregulated fishing in the Arctic Ocean. In December 2015, they met again and this time Japan, China, South Korea, the EU and Iceland were also included, and in the autumn of 2017 an agreement on unregulated fishing in Arctic waters was signed.[16] The point here is that it is the Arctic coastal states, Iceland and other major fishing nations that such issues are most relevant to. The Arctic Council is probably therefore less suitable as an appropriate venue for more in-depth cooperation (and ultimately an internationally binding agreement) around fisheries in the Arctic Ocean. Sweden and Finland do not have a large fishing fleet sailing in Arctic waters, and are therefore not legitimate actors when it comes to the management of these potential resources. Of course, the EU's Common Fisheries Policy also provides guidance on what roles they can play.

When it comes to questions about biodiversity and, ultimately, protected areas, issues of sovereignty are again the major challenge. Arctic coastal states want to be in the driver's seat in their respective economic zones and on the continental shelf. This is also reflected in the Arctic states' strategies for the region. As mentioned, it is only Sweden and Finland that have advocated for such regulation of the region. In recent years, however, PAME has worked on the kind of knowledge that should be the basis for establishing a protected area and has also served as a meeting place for discussing the issue, but it is the states that ultimately determine which areas shall be protected.[17] The Arctic Council's ability to inform decisions is more important than making decisions by negotiating an agreement between the eight member states. In an article published in 2016, the researcher and CAFF expert Christian Prip also said: "[T]he Arctic Council is still not ready to move from scientific cooperation and policy shaping to policy-making in the field of biodiversity."[18]

At the beginning and middle of the 2000s, as described in Chap. 1, the climate received increased attention in the work of the Arctic Council. An increased focus on emission reduction measures was a part of this development. In 2009, a task force that would look at short-lived climate drivers was established. One possible outcome of the establishment of such a temporary working group is recommendations for negotiating a binding international agreement under the auspices of the Council. This work has not resulted in such a recommendation. There may be several reasons for that, but a core argument is that despite the fact that emissions of short-lived climate drivers in polar areas are expected to have a greater climate impact in Arctic regions compared to emissions in non-polar regions, emission-limiting measures are essentially a global issue. It must and should therefore be addressed in other forums. Secondly, negotiating an agreement is much more politically and economically demanding. It does not mean that the Arctic Council does not have a role to play here. It can be noted, for example, that the EU, along with France, South Korea and Germany, have participated in an Arctic Council expert group on black carbon and methane (EGBCM). In this expert group, there is broad participation, where it may seem that the distinction between observers and permanent members is less than in other contexts. Nine of the observers (including the EU, France, South Korea and the UK) provided reports of measures taken in their own countries in 2015, several of these also provided updated figures published in the Black Carbon Report from

2017.[19] However, there is some way to go from such non-binding work to an agreement on reducing climate emissions in the Arctic. It is unlikely that the Arctic Council will go in that direction.

* * *

The decisions made in the Council are therefore not legally binding and some have argued that the Arctic Council is not an effective body, or put more bluntly it is an unnecessary and wasteful public body with no practical significance.[20] The knowledge produced in the working groups, and precise recommendations based on this knowledge production, are the main output of the Council. The recommendations of, for example, the Arctic Climate Impact Assessment (ACIA), the Arctic Marine Shipping Assessment (AMSA) and the Arctic Biodiversity Assessment (ABA) are only politically binding. National follow-up is therefore voluntary. The three agreements that have been signed are essentially exceptions that confirm the rule. They are primarily to do with improving coordination, which is of course important, but does not oblige the states and other commercial actors to regulate their activities (e.g. through emission limitations) in or of significance to the Arctic. The question is, therefore, whether the cooperation and knowledge produced and the recommendations that emerge from the working groups can have an impact on national priorities and international processes without negotiated guidelines and recommendations being legally binding.[21]

* * *

Broadly based reports prepared under the auspices of the Arctic Council have played a part in international and national policy design. There are also analyses that say something about whether and how this work has had an impact in international and national environmental and climate policy. There are many indications that “Arctic knowledge” has made a difference. Before we take a closer look at how and within which political areas Arctic Council knowledge has played a role, a few words about what is considered important for knowledge to make a difference.

There is a fairly broad scientific literature on the relationship between knowledge and policy. A starting point is often that in order for knowledge to “hold sway” in international decision-making arenas and in national administration, there must be scientific consensus on the nature of the problem and to some extent how it can be solved. A recurring example is the ozone negotiations, where the researchers agreed on cause and effect. However, there are also examples where the scientific consensus has greater challenges. Climate change is the crowning example here. In order to explain if it is successful or not, other variables must be considered. In the literature it is often argued that the degree of economic and political controversy surrounding a given topic is crucial. This is of course not surprising, but is nonetheless important to include in any analysis of the relationship between knowledge and policy. A third explanation that is often found is linked to how high a given challenge is on the public and political agenda. The degree of public and political attention affects whether knowledge is used. A fourth aspect that has empirically been proven to be of significance is how precise the knowledge (or any recommendation) is. A delimited field where the relationship between goals and means is clear will more easily catch the eye of governments and international processes and thus be implemented. One final factor is how the relationship between government and the research communities is organised and which actors are involved. Is this relationship more or less randomly organised, does it have clear formalised routines in terms of the researchers' access to relevant administrative bodies and international decision-making arenas, and are these characterised by a hierarchical structure or more or less equal lines of communication? Whether actors affected by political measures are involved in the process and whether there are systems for following up given projects and measures is also relevant. It is the work of the three "major" working groups, AMAP, CAFF and PAME, that are relevant in this context. Therefore, in this part of the chapter, I will say something about the effect of these working groups on international cooperation and national administration.[22]

* * *

The three working groups can be compared because they have carried out large ambitious assessments within their areas of responsibility. These have received attention both in and outside the Arctic. In addition, all three working groups were created under the precursor to the Arctic Council, the Artic Environmental Protection Strategy (AEPS), and became part of the Arctic Council structure in 1996, when the Council was formally established. Their reports typically contain recommendations for follow-up activities, and there is an overlap between their areas of responsibility, which mean that they often work together.

We can first look at AMAP. The overall picture shows that over two decades, AMAP has prepared a very weighty scientific basis for its activities. The regular assessments of, for example, persistent organic pollutants (POPs), mercury, radioactivity have a high scientific status. All scientific work in AMAP is also subject to peer review prior to publication. We can nevertheless make a distinction between mature areas and less mature areas. Radioactivity and POPs have a long history, and the condition assessments and research basis are clear enough to come up with well-founded recommendations that it is therefore difficult for political decision-makers to disregard. If we look at a new area, maritime litter, where it will not be unnatural that AMAP takes a central position over time, and where the research is at a fairly early stage, the scientific basis is more fragmented and uncertain. Of course, this has considerable significance for the opportunities for knowledge transfer and follow-up. In terms of climate, the second main pillar of AMAP, the picture is more complex. Despite broad scientific consensus that climate change is man-made, the scientific basis is considerably more fragmented. The various projections of the climate panel (IPCC) are a core example here. The sources of the problems are also more varied, something that makes the scientific work more complex and partly characterised by greater uncertainty. This, of course, has significance for how pointed the recommendations the working groups make can be. The area is at the same time characterised to a much greater degree by political and economic controversy. Put simply, tackling and regulating the pollutants in the Arctic is less demanding than a large-scale reorganisation of the world's energy structure. Another point is that scientific consensus is not necessarily static. For example, we see that industry conducts scientific work on the regulation of environmental pollutants and any new environmental pollutants that can be regulated. The Russian (and for that matter American) scepticism that climate change is man-made is another example.

Work with pollutants (including mercury and POPs) and climate still has a number of commonalities. AMAP's contribution within these two areas is to a large extent international, which has been demonstrated by, among other things, the cooperation with the UN Environment Program (UNEP).[23] We also find examples of national administrations using this knowledge in international processes. For example, Norway has actively used knowledge from AMAP about pollutants in the nomination of new POPs for the Stockholm Convention.[24] Here, work done under the auspices of the Arctic Council has been important for international regulations. Despite the fact that the work on pollutants has not led to legally binding agreements under the auspices of the Arctic Council, the work has received political blessings from the member states and has thus also been used directly in international negotiations, but individual states have also used this knowledge in their national environmental work.

Work that has a greater degree of national and regional impact is climate adaptation. The report "Adaptation Actions for a Changing Arctic" (AACA) from 2017 is a relevant example in this respect.[25] It is to a large extent a mapping of adaptation challenges and resilience in Arctic local communities and focuses less on specific measures. The Arctic states can therefore use this knowledge as a decision-making basis and less like a recipe for solving a problem. The area of adaptation, like the area of the climate in general, is cross-sectoral and it can also be argued that the knowledge base is more fragmented and sometimes unclear. This makes it difficult to track a direct effect on regulations, but we can see numerous instances of the knowledge being used to inform relevant actors about challenges and thereby indirectly also political decisions.[26]

A point in the extension of this is the degree of public and political attention around the areas AMAP is working on. An expectation will often be that the greater the political attention, the greater the chance of knowledge being recorded in national management processes. It is not given that it is like that. We can again look at AMAP's work with climate. Both at national and international level, climate is high on the political and public agenda. This means that national administrations and international negotiations are constantly exposed to new insights and complex political issues that increase the pressure on resources. At the same time, we see an Arctic Council context where reaching a consensus on how the knowledge base from the working groups is to be presented is politically sensitive. An example is from the climate negotiations in Bonn (2017), where various views on the role of the Arctic Council in these negotiations were debated. The outcome was that the work of the Arctic Council on the topics (based on AMAP knowledge) was presented in a side event during the negotiations. Anyway, the point is that within the area of climate the Arctic Council functions as an important knowledge provider, but clear and precise policy recommendations are more difficult to track.

Recurrent in the discussion on the use of AMAP knowledge is the degree of precision in the knowledge and the recommendations that often leap out of this. The documents that clearly show this dimension are the "summaries for policy makers", which are brief summaries of the comprehensive scientific reports. In the concluding parts of these scientific summaries, we often find more or less pointed conclusions and recommendations. Again, the pattern is the same. When it comes to pollutants, the recommendations are clear. It is quite clear which measures must/should be implemented. One example is the report “Arctic Pollution Issues 2015” where a very clear recommendation is that all the Arctic states must ratify the Stockholm Convention[27] and the Minimata Convention.[28] We also see that occasionally explicit recommendations are given regarding the regulation of individual pollutants. The two major reports on climate change in the Arctic, which follow up on the ACIA report, the SWIPA reports from 2011 and 2017 show another picture.[29] Here, the recommendations are broadly formulated, such as: “The Arctic States, permanent participants, and observers to The Arctic Council should individually and collectively lead global efforts for an early, ambitious and full implementation of The Paris COP21 Agreement, including efforts to reduce emissions of short-lived climate forcers.”[30] The work on short-lived climate forcers is nonetheless as previously shown more precise. Here the Arctic Council in 2017 approved a report on soot and methane with 12 concrete recommendations to reduce emissions. One example is recommendation 1c: “Reduce black carbon by stimulating the shift to alternative vehicle technologies and modes of transportation, and through transportation efficiency measures.” Another is recommendation 4a: “Avoid methane emissions by preventing the land-filling of organic waste.” It is still worth noting that these recommendations are not formulated by AMAP, but by an expert group.[31] It is another example of the tendency in recent years in the Arctic Council to appoint ad hoc groups, when more policy relevant recommendations are sought. The question we of course should ask ourselves is whether or not the work group’s efforts should move in the direction of such more precise recommendations. However, within the area of climate this is more complex. The role of AMAP is largely to provide condition assessments or to map new Arctic challenges. The input to, for example, the IPCC is largely purely scientific, where the purpose is to increase the knowledge base and therefore the decision-making basis for political actors, and not make final recommendations.

The main point here is that the degree of precision in the recommendations we find in AMAP’s reports varies. This is not surprising. Purely scientific measurements and condition assessments of individual pollutants will automatically lead to specific measures. We can also make a distinction between different types of recommendations, such as direct regulatory measures and the need for more research in a given area. However, it is again important to emphasise that none of these recommendations are legally binding.

We can, however, conclude that AMAP has had great success with its work on pollutants and heavy metals. When it comes to POPs, this activity has also led to the regulation of additional pollutants under the Stockholm Convention. This success is, however, not a product of strictly formalised routines between national administrations and international negotiations and the working group. When it comes to POPs and in that case pollutants in general, the conditions have facilitated the use of Arctic knowledge and this can be expected to continue. We can draw a similar picture when it comes to mercury. In addition to the factors that have been discussed above, the involvement of individuals (and in particular the previous secretariat leader Lars-Otto Reiersen), continuity in the leadership, and fairly stable funding have also probably been significant. On the other hand, the work on climate (including emission limitations and adaptation) has been more demanding. There are various explanations for this. Most importantly, the area of climate in itself is broader and requires a broader portfolio of measures across various sectors. It is therefore more demanding to formulate precise and concrete recommendations. The discussion above can be interpreted as it being all about context whether the power of the Council to succeed in the transfer of knowledge. It is a truth with modifications. Despite the fact that the degree of success is context-dependent, measures can be taken that will strengthen the relationship between the Arctic Council, international organisations and national administrations, something which I look at in Chap. 4.

* * *

The work of the other large scientific working group also tells us something about what the Arctic Council can contribute. CAFF largely lives up to the classic purpose of the Arctic Council to provide scientific knowledge about the Arctic environment—in this case, the state of the Arctic biological diversity and the factors affecting it. The impact outside the Arctic of CAFF's main products, the Arctic Ecosystem Monitoring Programme, the Circumpolar Biodiversity Monitoring Program (CBMP)[32] and the ABA,[33] are examples. The CBMP focuses on the Arctic's large ecosystems, fresh water, coast, sea and land, and has been given the status of being the Arctic arm of the global monitoring network the Group on Earth Observations Biodiversity Observation Network (GEO BON). The knowledge from the Arctic Council is therefore directly used in international processes. The CBMP's marine condition report contains a number of recommendations for follow-up monitoring tasks and for their coordination, which are highly relevant to nature surveillance in the individual Arctic states. In addition, CAFF has built up the Arctic Biodiversity Data Service (ABDS) data management system to facilitate access to information for researchers, decision-makers, local communities, international partners and so on. CAFF's work has scientific weight, and it is used and cited in an international context. It is included as regional contributions to global reports on the state of biodiversity. There is broad scientific consensus on the nature and scope of the challenges, but also that the solutions for them are extremely complex and dependent on a number of factors, which are largely beyond the control of the individual Arctic states. It therefore requires action outside the Arctic, but in other cases intergovernmental actions within the Arctic, such as the establishment of cross-border-protected areas. The latter requires binding agreements between one or more Arctic states on limitations in the use of areas under the state's national jurisdiction. Here, you immediately move into a politically sensitive area, and as previously mentioned, so far, there has been no political will in CAFF and the Arctic Council to lead the cooperation in such a binding direction that it affects the sovereign rights of states.

Biodiversity in the Arctic has public and political attention both among the Arctic States and internationally. CAFF's investigations identify shortcomings in nature surveillance and research and can therefore act as leverage for research and grants for research. Beyond that, the effects of CAFF's scientific investigations are difficult to measure at national level and must to some extent be designated as indirect in terms that CAFF and its technical expert groups are meeting-places for professionals, who take their experience from meeting with foreign colleagues home, allowing them to participate in their national work. Moreover, the work under CAFF provides "economies of scale" with opportunities for, for example, scientific expeditions with the participation of several Arctic states, opportunities that would be more difficult for the individual Arctic state to carry out alone. This of course also applies to the work of AMAP. In this sense, the work of Arctic Council's working groups has great significance.

As is the case generally for the Arctic Council's scientific investigations, CAFF's reports often contain recommendations for follow-up measures. However, this applies to a lesser extent to CAFF, and typically the CAFF recommendations are more general and less precise than from other working groups. Rarely does it state precisely whether it is the Arctic Council, CAFF, national administrations or others who are responsible for implementing the measures recommended. According to the CAFF action plan for the implementation of ABA, the ABA recommendations are as a rule directed at the Arctic Council as a whole and shall be implemented by both CAFF and other working groups, while only some of the recommendations will require action from national authorities, international organisations and other actors.

The point is that the overall topics of CAFF, biodiversity, ecosystems and ecosystem services, are less suited to concrete, precise guidelines for national action than topics under other working groups with more specific issues such as short-lived climate forcers, pollution control, oil spills and the like. This is due to CAFF's choice of a broad, cross-sectoral and ecosystem-based understanding of its role. That does not mean that the work done in this working group does not have an effect. The work is integrated in global work and will therefore be able to form the basis for strengthening international agreements on ecosystems and biodiversity.

As discussed above, the degree of precision is an important parameter for whether policies and recommendations have effect. It can therefore be argued that the often broad CAFF recommendations are less impactful than, for example, AMAP's more precise recommendations on the regulation of individual pollutants. The general form can lead to limited political and public awareness of the recommendations, and make it difficult to implement mechanisms for monitoring the national follow-up. Because CAFF's recommendations are neither formulated nor perceived as binding on the states, it is difficult to demonstrate their effect on national administrations. In those areas where CAFF is still working on specific and more tangible topics such as preparing action plans for endangered seabird species, it is easier to achieve a direct relevance for an individual country, and it is precisely in relation to CAFF's work on seabirds that the most direct effect is probably seen. The CAFF expert group for seabirds (CBird), for example, has drawn up international strategies and action plans for endangered species such as eider ducks, guillemots and ivory gulls.[34] These add up, among other things, up to national implementation plans. But the main picture is that CAFF has played an important role in contributing to international processes with, among other things, condition assessments of biodiversity and ecosystems in the Arctic. Knowledge built up by the Arctic Council is therefore important, but mainly for establishing a good scientific basis for decisions and not precise recommendations on measures in the member states of the Arctic Council.

* * *

The work of PAME can also tell us something about what the Arctic Council can do. Among the Arctic Council's working groups, PAME is the most policy-oriented. A particularly important task for PAME is to prepare proposals for specific preventive measures. This is expressed through the working group's policy recommendations, guidelines and action programme, which are aimed at both the Arctic Council and the eight member states. These products are intended to complement existing legal regulations, and common positions and policies are often inputted to the International Maritime Organisation (IMO), which, unlike the "soft" recommendations of the Arctic Council, establishes legally binding regulations and standards.

According to PAME's mandate, work on land-based pollution sources has been central to several of the working group's initial projects. In addition to the development of a regional action programme for land-based activities (though with Russia as the main focus), completed in 1998 and updated in 2009, PAME conducted a survey of the marine environment in the Arctic, its impact on land-based activities, as well as existing international instruments aimed at this type of pollution source. The report, which was published in 1996, pointed to four worrying forcers: POPs, heavy metals, radionuclides and oil. A status report including measures taken to follow up the report's recommendations was issued in 2002.[35]

In the last decade, it is, however, mainly on the "sea side" that PAME has made its mark, with topics such as shipping, an approach to ecosystem-based management and marine protection areas on the agenda. A number of projects have been carried out in these areas, one of the largest and most important works being the "Arctic Marine Shipping Assessment" (AMSA) (2009)—PAME's flagship and the first report ever to survey shipping in the north, its impact on residents of the region and the marine environment.[36] The report also addresses the demands that increased shipping place on the infrastructure in the Arctic. Other important examples are the reports from the project "Arctic Ocean Review Phase I and II" (2011 and 2013), which have contributed to compiling environmental status information, standards and regulations in the Arctic, with the overall aim of strengthening the governance of the marine environment through collaborative, coordination and integrated approaches to governance.[37] Also, the "Framework for a Pan-Arctic Network of Marine Protected Areas" (2015) is considered a significant contribution by PAME in terms of future cooperation, management and measures to protect vulnerable marine areas in the Arctic.[38] The report's vision is to establish a pan-Arctic network of marine protection areas that cross national borders, but which can be managed within the national legal frameworks of the Arctic states. With regard to marine protection areas, PAME has taken over the main focus of this work from CAFF, which was previously the most important arena within the Arctic Council for this.

PAME is an important arena for cooperation between Arctic states and for the exchange of knowledge and experience. The fact that PAME is run by expert groups is an important point here: objective and scientifically strong discussions with a focus on technical aspects characterise the work group. Among the experts, there is largely a common understanding of the problems and solutions, based on the facts available. The fact that the scientific consensus is relatively broad also gives a good starting point for knowledge transfer. However, there are examples of disagreements related to priorities and emphasis, as in the work on the AMSA report, where the balance between the environment and safety of individuals is referred to as being problematic.

In general, the thematic structure of PAME has received political and public attention in some Arctic states. Norway is a good example and has used knowledge built up in PAME actively. Norway is a major maritime power. Information and assessments made in the working group are important in order to raise standards among other member states, but also in national work. To a large extent, reports from PAME are considered scientifically solid, and knowledge from them has on several occasions been used by the national administration. Particularly useful is the method approach within ecosystem-based management, as well as workshops related to marine-protected areas. The latter because there has been less work on the topic in Norway. The administration therefore receives scientific "added value" through its participation in the working group.

Marine litter is another hot topic, where countries such as Canada and Norway are already active globally and nationally. Information from the ongoing PAME project "Desktop study of marine litter and microplastics in the Arctic" is useful in this respect, both for expanding the knowledge base for further work and for being at the forefront internationally. In general, we see that a presence in PAME provides scientific input which is often taken further in the national work—although this is not necessarily easy to track directly.

In some cases, this form of knowledge transfer has met resistance. The heavy oil project is one example. Originally, some member states attempted to promote a ban on the use of heavy oil in the Arctic, which was introduced in the area around Svalbard. However, it was not possible to get all the member states to agree on this measure, as the topic is politically controversial for several of the states in the Arctic Council—especially Russia. Heavy oil is used not only on ships, but also in homes. For Russia, which has a relatively large population in the north, a ban would therefore have major consequences. The final project has therefore consisted of investigating alternative fuels to heavy oil and assessing risks, resulting in several technical reports. The work nevertheless forms an important scientific basis, and a summary is currently being prepared which will be presented to the IMO. Ultimately, it is perhaps here that the most important transfer of knowledge takes place. PAME's work with the Polar Code and later on regional waste management for shipping also emphasises the working group as an important arena for forming common positions. In the long run, this can help to push international regulations in the desired direction.

Having said that—despite the fact that the Arctic Council does not have regulatory authority—one of PAME's most important tasks is to prepare policy recommendations for SAOs and ministers. From a knowledge transfer perspective, here, as previously mentioned, the precision of recommendations and reports is also an important dimension. Among PAME's products there is considerable variation on this point, but a trend in recent years is nevertheless that recommendations often point to measures that can be implemented to meet a given recommendation, as well as who is responsible for the follow-up (the Arctic Council, a specific working group or the states). With this in mind, precision does not appear to be crucial in the assessment of knowledge transfer between PAME and regulatory authorities in the Arctic states. One explanation is that knowledge transfer often takes place in the processes prior to negotiation and the issuance of policy recommendations—in workshops and discussions between scientific experts from the national administrations. In conclusion, it must again be emphasised that the recommendations and reports from PAME have a little impact on the policies of the Arctic States. Knowledge produced in the working group is nevertheless included in national assessments, and conversely, individual states input knowledge to PAME and the other member states. The most important role is again providing input to international processes, and in recent years the work with the IMO has in particular been central. Here, PAME has played an important role. The Arctic Council has thus set the agenda and given the politicians the opportunity to make hopefully well-considered decisions on the basis of a knowledge base established under the auspices of the Arctic Council.

* * *

We can therefore conclude this chapter on what the Arctic Council can do by saying that despite the fact that the eight member states have signed three international binding agreements in recent years, it is hard to envisage (and perhaps not appropriate to do so) a further legalisation of the Council's work. We must also remember that strictly speaking, it was not the Arctic Council that negotiated the agreements but the member states. This does not mean that the agreements do not matter. The most important thing at the time of writing is to conduct exercises, learn from them and make communication between the Arctic coastal states as easy as possible. We must also remember that agreements pertaining to international law are symbolically important, something that can be boasted about. They reinforce the legitimacy of the Arctic Council as such.

Some people have wanted to move further in this direction, that is, from what individuals would argue are more or less unclear political proposals and empty ambitions to legally binding resolutions. In what is referred to as the Kiruna vision from the ministerial meeting in 2013, this is explicit. The Arctic Council will move from "decision-shaping" to "decision-making".[39] This sounds basically reasonable. However, the Arctic Council is unlikely to move in such a direction. The two strongest actors (Russia and the US) will oppose it. It is not certain that it will be appropriate either. Former Norwegian SAO Karsten Klepsvik is said to have formulated it as follows: "It may be better to formulate clear, politically binding recommendations than vague legally binding agreements." A legal agreement is not worth more than its content. This is despite the fact that it may have a symbolic value. The point is that even though the decisions of the Arctic Council are not legally binding, they set the agenda and provide input to hopefully well-considered political decisions. At the same time, the network (with, e.g., NGOs and population groups) has a value. It establishes a legitimate venue for the exchange of views, which can and in several cases also has significance for decisions in the Arctic made at national and international level. It is not necessarily something to be sneezed at. In continuation of this, the Arctic Council is a community of interests where conflicting interests can also be aired without the goal being total agreement in the form of a diluted legally binding agreement. It is perhaps precisely this characteristic that explains why the Arctic Council has been a success in several areas. The main point, however, is that the Arctic Council will still primarily be a "decision-shaping" body and to a lesser extent a "decision-making" body.

If we look at the knowledge and the recommendations that emerge from the work in the working groups, these can have an effect. In this regard, it is also important to involve all relevant actors in processes that have regional consequences, but which must be handled in broader forums, such as emission reduction measures and standards for shipping. Here it is proactive, and here the Arctic Council has an important role to play. Based on this presentation of the Arctic Council's scope of action, where questions of what is politically and organisationally possible have been highlighted, the next step is to ask what should the Arctic Council do in order to continue to be perceived as the most important international cooperation forum in the Arctic?

Notes

1. The "neutrality" of Sweden and Finland can, of course, be discussed but there is not room for it in this book.
2. Classical texts within this tradition are John J. Mearsheimer (2001): The Tragedy of Great Power Politics. New York: W.W. Norton & Company and Kenneth Waltz (1979): Theory of International Politics. New York: McGraw-Hill.
3. Heather Conley and Matthew Melino (2016) "An Arctic Redesign Recommendations to Rejuvenate the Arctic Council".
4. The Ministry of Foreign Affairs in Denmark (2016), report: Danish Diplomacy and Defence in a Period of Conflict. Available from http://um.dk/en/foreign-policy/danish-defence-and-diplomacy-in-times-of-change/. Last visited 25 April 2019.
5. CBC News: On the 20th anniversary, experts say it's time to review the Arctic Council's mandate, 30/09/2016. Available from https://www.cbc.ca/news/canada/north/arctic-council-20-anniversary-lloyd-axworthy-1.3784144. Last visited 25 April 2019. It should also be noted that Lloyd Axworthy was central to the establishment of the Arctic Council in 1996.
6. Iceland is nevertheless in a special situation with sea territories north of the polar circle, but not land territory, and with open sea between it and the North Pole. It is perhaps therefore not very surprising that Iceland is somewhat sceptical about not being involved in these discussions.
7. Iceland, "A Parliamentary Resolution on Iceland's Arctic Policy", p. 4. Available from https://arcticiceland.is/en/island-og-nordhurslodhir/stefna-islands-i-malefnum-nordhurslodha. Last visited 25 April 2019.
8. Stokke, Olav S. "An inner circle in the Arctic Council?"

9. Purely anecdotally, it can also be mentioned that in the autumn of 2015 I gave a lecture on search and rescue in the Norwegian Ministry of Defence. At the time there were several in the audience who had spent a long time in the Ministry and had knowledge of search and rescue in the north, but had not heard of the agreement.
10. Agreement on Cooperation on Aeronautical and Maritime Search and Rescue in the Arctic (2011). Article 2.
11. Nevertheless, it is emphasised in the agreement that the delimitations in search and rescue regions are not linked to the boundaries between the states or their sovereignty, sovereign rights and jurisdiction. Ibid. Article 3.
12. For A closer analysis see Svein Vigeland Rottem (2013) "The Arctic Council and the Search and Rescue Agreement: the case of Norway."
13. Agreement on Cooperation on Marine Oil Pollution Preparedness and Response in the Arctic.
14. One of those who has written most insightfully about this is Olav Schram Stokke. For example, see Olav Schram Stokke (2013): "Regime Interplay in Arctic Shipping Governance: explaining Regional Niche Selection." International Environmental Agreements: Politics, Law and Economics, 13(1), pp. 65–85.
15. Svein Vigeland Rottem (2016): "The Arctic Council in Arctic Governance: the significance of the Oil Spill Agreement." In Linda Jakobson and Neil Melvin (eds.), The New Arctic Governance. Oxford: Oxford University Press, pp. 147–174.
16. The Norwegian Ministry of Foreign Affairs, Agreement on unregulated fishing in the Arctic Ocean. Available from https://www.regjeringen.no/en/aktuelt/agreement-on-unregulated-fishing-in-the-Arctic-ocean/id2580484/. Last visited 25 April 2019.
17. On the other hand, there is international work under the Convention on the Law of the Sea to establish protected areas outside national jurisdiction. For an analysis of this process see Christian Prip (2016).
18. "Towards a new legally binding instrument on the conservation and sustainable use of marine biodiversity of areas beyond national jurisdiction" Blog post Jebsen Centre. Available from https://site.uit.no/jclos/2016/10/21/towards-a-new-legally-binding-instrument-on-the-conservation-and-sustainable-use-of-marine-biodiversity-of-areas-beyond-national-jurisdiction/. Last visited 25 April 2019 and Christian Prip (2016): "The Arctic Council and biodiversity—need for a stronger management framework?", p. 43.
19. See "Expert Group on Black Carbon and Methane; Summary of Progress and Recommendations", available from https://oaarchive.arctic-council.org/handle/11374/1936. Last visited 25 April 2019.

20. This debate was particularly evident in connection with the commemoration of the Council's 20th anniversary in 2016. For example, see https://www.nrk.no/troms/arktisk-rad-feirer-20-ar_-_-har-en-lang-vei-a-ga-1.12766216. Last visited 25 April 2019.
21. Studies of whether international obligations (political and legal) have an effect on national priorities (administrative measures or budget allocations) are one of the most important fields in analyses of international politics. One of the most frequently quoted contributions in an Arctic context is Olav S. Stokke & Geir Hønneland (eds.) (2007) International Cooperation and Arctic Governance: Regime Effectiveness and Arctic Governance, London & New York: Routledge.
22. This part of the chapter rests on the FNI report The Arctic Council in Norwegian environmental management from 2018 written by Svein Vigeland Rottem, Christian Prip and Ida Folkestad Soltvedt. A big thank you to Christian Prip and Ida Folkestad Soltvedt for allowing me to use key findings in the report in the book.
23. AMAP/UNEP, 2013. Technical Background Report for the Global Mercury Assessment. Available from https://www.amap.no/documents/doc/technical-background-report-for-the-global-mercury-assessment-2013/848. Last visited 25 April 2019.
24. Svein Vigeland Rottem (2017): "The Use of Arctic Science: POPs, Norway and the Stockholm Convention".
25. AMAP, Adaptation Actions for a Changing Arctic: Perspectives from the Barents Area (2017). Available from https://www.amap.no/documents/doc/adaptation-actions-for-a-changing-arctic-perspectives-from-the-barents-area/1604. Last visited 25 April 2019.
26. See Svein Vigeland Rottem, Ida Folkestad Soltvedt og Christian Prip (2018) "The Arctic Council in Norwegian environmental management." For a more detailed analysis of such conditions.
27. Stockholm Convention on Persistent Organic Pollutants (POPs) was signed in May 2001 and came into force on 17 May 2004. The purpose of the convention is to regulate the use of POPs in industry and agriculture. The substances are mainly man-made and can have a significant negative impact on health and the environment. For more information see http://www.pops.int/. Last visited 25 April 2019.
28. The Minamata convention, which was signed in October 2013, is a global agreement to regulate and reduce mercury emissions. Mercury is like POPs, and a long-range pollutant and regulation is therefore also of significance to Arctic conditions. For more information see http://www.mercuryconvention.org/. Last visited 25 April 2019.

29. AMAP, Arctic Pollution Issues 2015. Summary for policy-makers (2015). Available from https://www.amap.no/documents/doc/summary-for-policy-makers-arctic-pollution-issues-2015/1195 AMAP, Snow, Water, Ice, Permafrost in the Arctic (2011). Available from https://www.amap.no/documents/doc/snow-water-ice-and-perma-%20frost-in-the-arctic-swipa-climate-change-and-the-cryosphere/743. Last visited 25 April 2019and AMAP, Snow, Water, Ice, Permafrost in the Arctic (2017). Available from https://www.amap.no/documents/doc/snow-%20water-ice-and-permafrost-in-the-arctic-swipa-2017/1610. Last visited 25 April 2019.
30. AMAP, SWIPA. Summary for Policy-makers (2017). Available from https://www.amap.no/documents/doc/snow-water-ice-and-perma-frost.-%20summary-for-policy-makers/1532. Last visited 25 April 2019.
31. Expert Group on Black Carbon and Methane, Expert Group on Black Carbon and Methane: Summary of progress and recommendations 2017.
32. CAFF, CBMP (2018). Available from https://www.caff.is/monitoring. Last visited 25 April 2019.
33. CAFF, Arctic Biodiversity Assessment (2013). Available from https://www.caff.is/assessment-series/233-arctic-biodiversity-assessment-2013. Last visited 25 April 2019.
34. For more information, see https://www.caff.is/seabirds-cbird. Last visited 25 April 2019.
35. PAME, PAME report on the status of the 1996 recommendations (2002). Available from https://www.pame.is/index.php/document-library/pame-reports-new/pame-working-group-meeting-reports. Last visited 25 April 2019.
36. PAME, Arctic Marine Shipping Assessment (2009). Available from https://oaarchive.arctic-council.org/handle/11374/54. Last visited 25 April 2019.
37. PAME, Arctic Ocean Review phase I and II (2011, 2013). Available from https://arctic-council.org/index.php/en/our-work2/8-news-and-%20events/327-arctic-ocean-review-project-aor. Last visited 25 April 2019.
38. PAME, Framework for a Pan-Arctic Network of Marine Protected Areas (2015). Available from https://www.pame.is/index.php/projects/marine-protected-areas. Last visited 25 April 2019.
39. Arctic Council, Vision for the Arctic (2013). Available from https://oaarchive.arctic-council.org/handle/11374/287. Last visited 25 April 2019.

CHAPTER 4

Challenges and Recommendations

Abstract The key question in this chapter is how best to utilise the knowledge generated within the framework of the Council. One proposal is to draft a clearer vision for the Council by, for example, creating an Arctic Summit. Regarding structural changes and closer coordination, it is argued in this chapter that the Council should appoint an expert panel to discuss and recommend steps to improve coordination, identify overlaps, and propose, if necessary, a reorganisation of the structure of the WGs. The final proposal is to locate an SAO meeting in a capital city and invite relevant organisations to attend and take part in an Arctic Week. This proposal is closely tied to the idea of a clearer vision, but also to the practical challenge of the growing frequency of travel and capacity constraints in Arctic venues.

Keywords Arctic Council • Vision • Structure • Participation

With the Arctic Council's history, structure and tasks fresh in mind, it is natural to say something about what the Arctic Council should do in order to continue to be perceived as the place to be when the future of the Arctic is being discussed. I would also like to venture to make some recommendations.[1] The aim of previous chapters was also to say something about the room for manoeuvre and the opportunities this cooperation has in the legal and political landscape in which it moves. It was necessary to understand what kind of potential the Arctic Council has and what proposals for change are politically possible. The role and structure of the Arctic Council

S. V. Rottem, *The Arctic Council*,
https://doi.org/10.1007/978-981-13-9290-0_4

in general and the working groups in particular have been the subject of debate in a number of Arctic Council meetings, consultancy studies and the scientific literature. A brief introduction to important contributions is appropriate.

In 2001, Pekka Haavisto, then at the Finnish Institute of Foreign Affairs, conducted an analysis on the structure of the Arctic Council.[2] The background was that during the Finnish chairmanship from 2000 to 2002, Finland commissioned an evaluation to be presented at the SAO meeting in Rovaniemi in June 2001. This was discussed there and the final report was published in late June 2001 and circulated among member states, permanent participants, working groups and observers. Haavisto's work is still relevant and is often highlighted as one of the most important contributions when it comes to discussions about the Council's challenges and opportunities. The report is ambitious and sets up both short-term and long-term options for the structure of the Arctic Council. It is broadly based and addresses a number of challenges, including overlap between the working groups, knowledge gaps, internal competition between the working groups, economic issues and cost efficiency. The most interesting discussion deals with the reorganisation of the working group structure. One suggestion that is discussed is to reduce the number of working groups to two: a monitoring group and an implementation group. In such a structure it is thought that AMAP should function as the monitoring group and PAME as the implementation group. An alluring thought, but probably both practically and politically impossible. A main argument against such a total restructuring was and remains that this will weaken the work being done in the working groups. The data that is collected, the people that are involved, the motivation of the individual scientists and for that matter also the member states will be affected.

The structure proposed by Haavisto is therefore somewhat less extensive. He proposes creating four working groups: brown, blue, green and rainbow. Of course, it can be argued that such colour coding may appear as being artistic, but the points from the report are nevertheless well thought out. The proposal is that AMAP and ACAP are merged into the brown group.[3] Here, surveillance of the Arctic environment will be the central function, but the incorporation of ACAP will also give the group an implementation function. The group will therefore work towards creating a prioritised and updated list of projects that take action against pollution. The idea is thus to combine monitoring and implementation. The blue group will undertake what is now PAME and EPPR's mandate, which is to come up with measures to strengthen the work specifically against marine pollution (including oil spill protection), both with regard to preventive measures and cooperation regarding any response in the event of an accident. CAFF will, like the green-group, work with living resources and biodiversity. Not unlike the role the working group currently has. In this regard, Haavisto also discusses a merger of AMAP and CAFF, where AMAP takes over the monitoring function CAFF has in relation to biodiversity. Haavisto still believes that this is not the way to go, cf. the arguments above. In the fourth and final group, the rainbow group, the SDWG will continue to work with sustainable development in the Arctic.

In the aftermath of this rather critical evaluation, the SAOs prepared a report for the ministerial meeting in Inari, Finland, in 2002. It took the work done by Haavisto as the basis for the report. There was no major restructuring, and the SAOs also presented a list of fairly general recommendations for the involvement of all working groups with sustainable development, the further inclusion of the observers in the working groups, a requirement for the stricter prioritisation of projects and more clarity in the mandates of the working groups (and thus also the coordination between them).[4] The ministers at the meeting approved the SAOs' report, and the work in the Arctic Council continued as before.

In 2008, the Norwegian chairmanship (2006–2009) again put the overall question of the Council's structure on the agenda. At an SAO meeting in Kautokeino in November 2008, the Norwegian chairmanship presented a report in which the following topics were addressed: funding, priorities, dissemination and coordination and communication between the working groups.[5] Funding is highlighted as a core challenge, which also potentially has consequences for project quality. It also appears that funding is closely connected to prioritisation. One recommendation is that the Council to a greater extent makes longer plans that make prioritising easier. When it comes to coordination, one benefit that is highlighted is that the head of the SAOs has regular meetings with the working group leaders. The last point is also connected to dissemination, where involvement and cooperation with other actors active in the Arctic must be strengthened, including the observers.

In 2012, Russia, Norway, Sweden, Denmark and the US agreed to carry out an inspection of the work done by the national authorities with the Arctic Council, work that was led and coordinated by Norway and Russia. At the SAO meeting in Anchorage from 21 to 22 October 2015, a summary report was presented. This is the most thoroughly prepared evaluation of the Arctic Council ever.[6] Key findings in this multilateral audit were: (1) changes in the Arctic have increased the importance of international cooperation in the Arctic; (2) the Arctic Council has contributed to increased collaboration, management tools and scientific knowledge; (3) the Arctic Council faces important challenges related to its organisational structure, lack of clearer priorities, funding of its work and ensuring effective implementation of recommendations; and (4) indigenous groups make important contributions to the Arctic Council, but face a number of challenges when it comes to participation.

The audit underlines challenges associated with the increasing number of projects under the auspices of the Council. The Council has expanded its area of impact and increased its workload considerably. The challenge is therefore to fund this work and follow up on the recommendations that emerge. An interesting observation is that there is a great deal of variation between the working groups with regard to perceptions about the extent of the problem regarding structure and overlapping mandates. For example, representatives from AMAP, EPPR and SDWG see the overlap as a problem, but representatives from ACAP, PAME and CAFF perceive the current structure as adequate.[7]

On the other hand, in the multilateral audit, it is pointed out that communication between the working groups has improved, something has therefore happened between when Haavisto conducted his work and the audit's work. The secretariat and the chairs of the working groups are meeting more often to share information. The establishment of the permanent secretariat in Tromsø may also strengthen coordination between the working groups and provide support to the working groups in their dissemination work. There are no thorough analyses of whether the secretariat has managed to fill this role, but things indicate that it still has some way to go.[8] The bringing together of three of the working groups (AMAP, EPPR and ACAP) in Tromsø means that there is also a reduction in the physical barriers for contact. In connection with this topic, the multilateral audit also shows that the increased use of "task forces" can lead to resources being withdrawn from the working group's projects. On the other hand, it is nevertheless emphasised that these time-limited working groups can provide complementary expertise.

Traditionally, the working groups will often identify their projects and then get approval at ministerial and SAO level. In this way, the working groups have a fairly large degree of autonomy. This can be a challenge. On the other hand, the development of the "tracking tool for Arctic Council deliverables and ongoing work" has increased the possibilities for insight into ongoing projects.[9] This is an updated overview of progression and contributions from projects under the auspices of the Arctic Council. A recurring challenge, however, is that the Council lacks mechanisms for prioritising the work done in the working groups and task forces. This must be seen in the context of the fact that the Council lacks a clearer strategy for its work beyond more general visions such as those formulated at the meeting in Kiruna in 2013. Which in turn leads to finding funding and the right expertise in the given fields becoming a problem. According to the audit, this problem must be followed up, in particular on the basis of the ever-increasing project portfolio. In general, stable funding is also a problem. A relevant tool here is the Project Support Instrument (PSI). A fairly new tool, where the states can voluntarily facilitate the funding of projects under the auspices of the Council.[10] In 2018, this measure was under external evaluation and hopefully this will give a better idea of how successful PSI is.

The conclusion of the audit is therefore that the recommendations of the Council are often broad and general and therefore difficult to implement and if implemented difficult to check. Here the link to the national level is also crucial. The individual states do not have good routines for following up work done in the Arctic Council.[11] The audit therefore points out a number of challenges in the work of the Arctic Council. A meeting that was held in Tromsø in September 2015, where four of the working groups participated, is nevertheless an example that we are seeing an increasing awareness of the topics looked at in the audit. This was the first attempt at convening a large coordination meeting. Another is the meeting held in Reykjavik in December 2015, where several leaders of the working groups met. Therefore, in the autumn of 2015 an active debate was taking place. The reports from the national supervisors and the multilateral audit were central to this, but the then head of the SAOs, David Balton, was also active. His role as a driving force for better coordination and organisation of the Council during the US chairmanship period from 2015 to 2017 was important. The latest addition in this constantly recurring work on how to improve the Arctic Council is the ongoing work between the SAOs on creating a strategy for the Arctic Council. At the time of writing, this has not been completed, but will be presented (or further discussed) at the ministerial meeting in Iceland in 2019.

We also find some academic contributions that have looked at this topic. An interesting, important and frequently quoted contribution is Paula Kankaanpää and Oran Young from 2012.[12] They start with a fairly positive analysis of the Council's work and effectiveness. The influence the Council's work has had nationally and especially internationally has exceeded the expectations of many, they say. But it is important to examine what measures can be taken to ensure and improve the effectiveness of the Arctic Council in the coming years. The article is based on a survey that was sent to people (primarily researchers and administrations) familiar with the Council and its work. In conclusion, Kankaanpää and Young highlight good results in terms of knowledge generation and the setting of the political agenda. However, potential for improvement is also discussed. They make a distinction between internal conditions and external problems. On internal matters, their focus is on the structure of the working groups and division of work between them and the task forces. When it comes to external issues, it is important, according to the authors, to engage regional and local groups, together with large non-Arctic states, including the new observers.

* * *

The point of this brief presentation of key contributions discussing the Arctic Council's challenges is to show that much remains the same and change is not easy. The effectiveness of the work done in the Arctic Council has been the subject of great attention at national, working group, SAO and ministerial level as well as in independent research. And by just a quick glance over the minutes of SAO and ministerial meetings from recent years, we can see that having a clearer vision for the cooperation, better coordination and increasing the involvement of observers have been recurring issues and are increasingly being discussed.

Thus, what this brief overview tells us is that there is a long history of critical discussion around whether the Council is functioning optimally. A debate that goes in cycles, but perhaps somewhat more frequently in recent years. The challenges have become more pressing with more and more observers and projects under the auspices of the Council. However, it is difficult to find common ground. There is no quick fix and it will take time to find the optimal institutional structure for the Arctic Council, if this is possible at all. At the same time, in recent years there has been more and more awareness of these challenges and much work has been done, among other things, on strengthening coordination and information sharing between the working groups, both formally and informally. An important recognition, however, is that it is necessary to balance what is politically possible with what seems rational from individual actors (see also Chaps. 2 and 3), be it a member country, or a representative of a working group or an indigenous group. Since each working group and each state is different, the function, role and focus of the Council's work will always be debated. There is no such thing as one Arctic, and so there will always be disputes between the actors. It should nevertheless be an important ambition to create a common vision/strategy for the council, strengthen coordination and establish permanent arenas for participation for a wide range of actors with legitimate interests in the Arctic.

* * *

One of the most important hallmarks of the Arctic cooperation is the revolving chairmanship. Such an arrangement has two important characteristics. Firstly, it provides member states the opportunity to set the agenda in their chairmanship period and thereby potentially initiate political processes that are considered particularly important by the country in question. By extension, each state gets ownership of the direction in which the Arctic Council will go. This also leads to greater involvement from the member states. On the other hand, a disadvantage of the rotating chairmanship is that continuity is lost, and there is a danger that states that do not have the chairmanship in a given period will be less inclined to follow up on specified political priorities. In the worst case, work in the Arctic Council will be given lower priority, as the Arctic agenda is not sufficiently anchored in the States. It may on the other hand it can be argued that the Arctic Council, as a consensus body, will to a lesser extent face this problem. In any case, all the time the Arctic Council is not an international organisation with the possibility and the ability to sanction states that do not comply with political decisions, this will be a recurring challenge. The question, then, is how to increase the legitimacy of the work in the member states. A legitimacy that in turn will lead to increased political attention, also during periods when one does not hold the chairmanship. The Finnish Arctic Strategy from 2013[13] and several statements from the Arctic Parliamentary Assembly[14] introduced a possible initiative. The proposal was for an Arctic Summit.[15] Such a summit could be held, for example, every four years. In this connection, it would be natural to work out an agreed vision for which direction the Arctic cooperation should go in and which topics should receive extra attention. Such a declaration would again provide guidelines for the work at the SAO level and at the working group level. This would again put a further pressure on the organs of the Arctic Council to coordinate and focus on measures that follow the overall vision formulated at the highest political level.

The challenge of such a recommendation is, of course, that focus at a political level will be on what is politically opportune for a limited period. It will, however, still strengthen the work within given political priority areas. On the other hand, it may mean that the work in some working groups will not receive the attention wanted. This does not necessarily mean that the central Arctic work will be forgotten, but rather during given periods the Arctic Council should wield more political weight in certain areas. One consequence is that other important Arctic questions of a more regional, national and local nature must to a greater extent be prioritised nationally and to a lesser extent occupy the daily activities at the Arctic Council level, and especially the SAO level. This is why pan-Arctic challenges will be prioritised, which in turn will ensure broad support for such an overarching vision and hopefully give all states ownership of it. In preparing such an overall strategy, greater inclusion of the sector ministries in the Arctic states will be natural. Challenges linked to the involvement of sector ministries have been recurring in several previous evaluations and most clearly in the work from the multilateral audit in 2015.[16] Such honing of the work will also lead to better coordination and ultimately implementation (see also below). A key realisation that underlies such a recommendation is the growing portfolio of the Council, where as previously described more and more actors are joining the cooperation, and the list of projects is getting longer and longer. The way the Arctic Council is currently structured, including a small secretariat, means that there are limits to how many measures/projects/fields of work can be maintained without this being at the expense of the possibility of real implementation and a strong political agenda function.

If we look at the individual working groups, we can see that such strategy work takes place. One example is CAFF's strategy, Actions for Arctic Biodiversity 2013–2021. Implementing the recommendations of the Arctic Biodiversity Assessment.[17] This formulates a set of goals and means that can be updated and revised every second year. Where there are no specific reasons for clarifying a longer strategy, all the work groups should continue to formulate strategies over a longer period of time. In the smaller working groups (ACAP, SDWG and EPPR) such a working group strategy may initially seem somewhat demanding, although it is the practice of EPPR to update its strategy every five years through the SAOs. However, the work in these groups will probably be closer to the overall vision. ACAP, SDWG EPPR's role will therefore to a greater extent be to function as functional units for the practical follow-up of specific measures formulated at the ministerial level. An obvious example is the role of EPPR in the follow-up of the two internationally binding agreements negotiated between the member states.

A clearer vision/strategy for the work being done in the Arctic Council, both at political level and working group level, will facilitate increased continuity and therefore a more structured and coordinated approach to selected political priorities. This therefore becomes a top down initiative, but with the main focus on substance not organisation. Nevertheless, the political priorities (vision) will also have consequences for the structure and coordination of working groups.

* * *

With the establishment of the permanent secretariat in Tromsø in 2013, the capacity for knowledge transfer, both to the international public and within the Arctic Council, has increased. The establishment of a permanent secretariat can, to a greater extent, make it easier to coordinate between different bodies in the Arctic cooperation. It is not surprising that we see a certain overlap. One example is that virtually all groups have a maritime focus and that pollution is part of the agenda. There is a broad awareness of this topic in the work of the Arctic Council. This can be seen, among other things, in formal and informal arenas for the coordination of procedures and potential overlap between projects. On the other hand, there are several conflicting perceptions about the extent of the challenge, which, among other things are made evident in the report from the multilateral audit. An obvious question is whether this overlap should lead to a change in the working group structure and a potential reduction in the number of working groups. I do not want to make any recommendations regarding reduction or restructuring. That does not mean that there should be a critical debate on overlapping mandates and measures for better coordination. If we look at previous evaluations, there are some issues that have received extra attention. It is therefore natural to briefly summarise these. This does not mean that you can find similar challenges in and between other working groups. A recurrent issue is the relationship between CAFF and AMAP. In the critical review of the working group structure from 2001, among other things, a merger of these two working groups was discussed.[18] The argument is that overlap exists and it may seem somewhat constructed to have two large scientific groups making condition assessments of Arctic nature. The question is whether in the current situation such a merger should be recommended. A top-down merger of two of the most central working groups would be politically demanding. A hallmark of the Arctic Council's work is that the working groups are spread among the member states. Such a spread with, for example, CAFF in Iceland and AMAP in Norway, distributes ownership of the work done in the Council. A merger will require centralisation and would be extremely demanding to undertake. Therefore, there should be strong arguments for such a measure being necessary. This is not to say that there should not be a discussion about whether such a measure may strengthen cooperation, but a danger is that established networks will deteriorate and motivation among individuals and member states will be harmed. It can also be noted that CAFF's clear focus on biodiversity is unique, and the same applies to AMAP's work on pollutants. It is clear that these fields of work cannot be understood independently, but any reorganisation must be clearly rooted in substantial challenges and not structural considerations as such.

In previous evaluations of the Arctic Council, SDWG also receives repeated attention. The question is whether all the working groups should have a clear sustainability element or whether questions concerning sustainable development in the Arctic should only be anchored in one working group, but with a strong connection to the newly created Arctic Economic Council (AEC). The AEC is intended as an arena for the involvement of commercial actors in Arctic cooperation, and thus cannot be seen independently of the SDWG which has just sustainable economic development as its main pillar. SDWG is the working group that most clearly differs from the others, among other things by its defined focus on sustainable development. However, there is a scarcity of proposals for how SDWG could be integrated into other work groups or possibly strengthened. A recurring challenge is that SDWG becomes too local in its area of impact. Topics that affect the entire Arctic population must be prioritised to a greater extent. Relevant fields are education, health and other socio-economic commonalities in the population in Arctic areas, this is not to say that this is not being done, but there is a potential for improvement.

The relationship between PAME and EPPR is another recurring issue. The distribution of mandates between PAME and EPPR is unclear. This is particularly the case within the area of pollution, where overlap can be seen. Basically, EPPR is more technical while PAME is more policy-oriented. This is irrespective of the potential for improvement and further clarifications of the mandate.

In recent years we have also seen that several task forces have been established. The creation of these gives some topics a higher place on the agenda. Several have expressed concern that these time-limited expert groups will compete against the working groups when it comes to funding, among other things. The practice of establishing more task forces and expert groups also expands the portfolio (both structure and theme), which can be challenging in terms of coordination and possible overlap between projects. At the same time, the recommendations and any agreements negotiated within these expert groups and task forces must be followed up. Again, EPPR is an obvious example, as the group has a central role in following up the 2011 and 2013 agreements, but the process that led to an agreement was led by a task force. The question is whether it was necessary to establish a separate expert group when the EPPR could have been a natural venue for the preparation of the agreement. One should strive to incorporate new political initiatives into the existing working group structure, where there are no obvious reasons for anything else.

In a broader and inclusive debate about a merger, the discussion in the review from 2001 is a reasonable starting point. Any such restructuring should in any case be rooted in the entire Arctic Council. A recommendation is therefore to establish an expert group that deals with the question of coordination and possible restructuring. Such a recommendation can of course be claimed to be a paradox. The last thing we need is perhaps yet another expert group? But here, the goal sanctifies the means. Here, too, the relationship between the purely scientific function and the implementation function should be high on the agenda. A clear mandate is also required.

Such a debate must involve all relevant actors (the indigenous groups, SAOs and working groups). Such a discussion could lead to a changed working group structure, but that is not a necessity. It is evident that previous proposals for restructuring and reorganisation have not been well enough rooted in the Arctic Council. This may go some way to explaining why we are seeing ever-recurring discussions on these questions without seeing any overall changes. Not to say that these debates have not brought with them a clearer awareness of coordination challenges, which have come to light, among other ways, through facilitating cooperation to a greater extent across the working groups and in establishing the permanent secretariat in Tromsø. Now there is momentum to go one step further in this process.

In continuation of the debate on coordination at the Arctic Council level, the 2015 multilateral audit emphasises that the main challenge in the Arctic Council is most often national coordination. However, there are considerable variations between the eight member states. The national challenges are different, so to make recommendations one has to make studies of each individual country. However, some issues are common: the transfer of knowledge when replacing SAOs should be strengthened, and clearer inclusion of the sector ministry and the establishment of national coordination arenas is a necessity, as well as a more conscious focus on the implementation of recommendations and guidelines from the Arctic Council. All the member states have potential for improvement in this respect. By facilitating increased and coordinated participation of relevant actors, it is nevertheless possible to facilitate better coordination at the Arctic Council level as well.

* * *

A central debate in the Arctic Council in recent years has been linked to the inclusion of several observers. Questions about participation have been frequently discussed. However, the discussion on participation cannot be limited to addressing the new observers from Kiruna 2013. More longstanding observers and indigenous groups must also be included, as well as national sectoral ministries and other regional and international arenas for cooperation.

A recurring challenge in the Arctic Council is the frequency of travel and an increasing number of meetings. To meet this challenge, one option is to hold one of the two annual SAO meetings in a capital. This could at the same time lower the threshold for participation by actors with legitimate interests, including the observers. In connection with such an SAO meeting the working groups will also be able to conduct workshops with relevance to the agenda. Working group meetings are already held ahead of SAO meetings, but a strengthening of this practice will allow for further coordination.[19] This will apply both to the design of new projects and the implementation of measures already put into effect. One recommendation is therefore to hold an annual "Arctic Week" in the chairmanship's capital. Now it can of course be noted that the last thing the Arctic needs is a new seminar on challenges and opportunities in the region.[20] It is therefore important to emphasise the function of such a meeting place, namely to support the work of the Council, and not to simply function as a network for Arctic brainstorming. Such a proposal is also rooted in the role of the Arctic Council in the legal and political governance of the Arctic. The main function of the Arctic Council and in particular the working groups is to input knowledge to national (e.g. various sector ministries) and international (e.g. the IMO, the UNFCCC, the Stockholm Convention) decision arenas. By facilitating meetings, knowledge transfer will be strengthened. It can be argued that such a practice will create a clearer awareness of the Arctic Council's role in what is often referred to as "Arctic governance".[21] At the same time it is important to maintain a clear footprint in the Arctic region, including through the lion's share of Council meetings being held in Arctic areas. In connection with a broader participation of observers, indigenous groups will also be able to raise an awareness among relevant actors of their central role in the Arctic. A clearer vision in the Arctic cooperation may also increase political attention, which in turn can facilitate increased funding.

One last point that deserves some extra attention here is the role of the observers—and especially in the working groups. It is generally accepted that the inclusion of new observers has been successful. Following the decision in Kiruna in 2013, we saw that the role of the observer appeared to be somewhat unclear. The prepared guidelines were to general, which also meant that at the SAO meeting in Anchorage, Alaska, in October 2015, an addition to the observer manual was adopted. However, efforts to coordinate the procedures for including observers in the working groups should not be terminated and not be dependent on the priorities of a particular chairmanship. One option is to look more closely at whether this can be clarified in the guidelines for the Arctic Council's work. In general, however, we see that the observers have a central role, and despite varying participation in the working groups, we have seen a positive shift, where the observers are providing case-specific input to the work. We have therefore seen a shift from a more general political awareness of the Arctic to a clearer desire to contribute scientific resources into the working groups. One initiative that has been put forward by the observer countries is that the dialogue between the rotating chairmanship of the Arctic

Council and the observer countries should follow the Warsaw format launched in 2010. That is, in meetings between the observer states and the chairmanship where information and views can be exchanged.

* * *

To sum up this chapter, the proposal to prepare a clearer vision for the Arctic Council, for example through the establishment of an Arctic Summit, is based on the continually recurring point that continuity and a lack of a clear strategy is a key challenge. It is not an argument for the Arctic Council to go from being an international forum to an international organisation.

Furthermore, the perceptions of any challenges faced by the Arctic Council and their magnitude are cautiously different. It will therefore be unnatural to make clear recommendations here. A unified discussion about what challenges are being faced and how to solve them is still required. The Arctic Council should therefore establish an expert group that can discuss as well as come up with proposals to improve coordination, identify overlaps between the working groups and recommend a reorganisation of the structures of the working groups. Furthermore, several people have pointed out that it is most often at the national level that coordination is most demanding. This is also expressed in the work of the multilateral audit. To have a look at its findings again is therefore also natural.[22]

Lastly, a discussion about participation is closely linked to the debate on the Arctic Council's position in Arctic governance. The Council's current role is, as mentioned, mainly as a provider of knowledge that prepares guidelines and recommendations that are more or less politically binding, but not legally binding. If this work is to be strengthened, the interaction with international organisations with decision-making authority is crucial. Available venues for cooperation should therefore be a goal. One recommendation is to hold an annual SAO meeting in a capital, and invite relevant actors to an Arctic Week. Such a proposal is again closely linked to the idea of a clearer vision, but also to the practical challenges associated with increased travel frequency. Such a process will of course be demanding and is therefore also an argument for a further strengthening of the permanent secretariat in Tromsø. The proposals that are presented here can be criticised for being too complex. Among other things, this will require an overall discussion of the Arctic Council's priorities in an Arctic in constant change. It appears, however, to be required.

Notes

1. The starting point for these recommendations and discussions around them is largely derived from interviews and talks with people with long-term involvement in the Arctic Council and is largely based on the FNI report "Arctic Council: Vision, Structure and Participation" (2016), author Svein Vigeland Rottem. For a list of interviewees see https://www.fni.no/getfile.php/131762-1469869078/Filer/Publikasjoner/FNI-R0416.pdf. Last visited 25 April 2019. Overlapping themes are also dealt with here: Svein Vigeland Rottem, Ida Folkestad Soltvedt and Christian Prip (2018) "Arctic Council in National Administration". A list of interviewees is also attached. It can of course be noted that for methodological reasons, such disguising of interviewees is not good. My experience of analysing the opportunities and challenges of the Arctic Council is that few people are willing to talk unless they are anonymised. One may then ask why this is. It may seem that internal and external criticism is not always well received in the Arctic Council. It is a challenge in itself.
2. Pekka Haavisto, Review of the Arctic Council Structures (2001). Available from https://oaarchive.arctic-council.org/bitstream/handle/11374/449/ACSAO-FI01_6_AC_Structure_final.pdf?sequence=1. Last visited 25 April 2019.
3. ACAP was not formally a working group at the time, but functioned as an independent entity.
4. The Arctic Council (2002), SAO's Report to Ministers on the Review of the Arctic Council Structure. Available from https://oaarchive.arctic-council.org/handle/11374/515. Last visited 25 April 2019.
5. The Arctic Council (2008), Improving Effectiveness and Efficiency of the Arctic Council. Norwegian Chairmanship. Available from https://oaarchive.arctic-council.org/handle/11374/1668. Last visited 25 April 2019.
6. The Arctic Council (2015), Joint Memorandum of a multilateral Audit on the Arctic states' National Authorities' Work with the Arctic Council. Available from https://oaarchive.arctic-council.org/handle/11374/1526. Last visited 25 April 2019.
7. The number of interviewees in the audit's work is nevertheless limited, and it can therefore be said that these assessments are largely dependent on the perceptions of individuals. In 2016, however, I worked on a report that addressed overlapping topics, and here I also found great variations between the working groups, which strengthens the audit's assertion about there being different perceptions of what challenges the Arctic Council is facing.
8. For an analysis of the role of the secretariat in relation to the SDWG working group, see Ida Folkestad Soltvedt and Svein Vigeland Rottem (2017)

"Challenges of the Arctic Council's Sustainable Development Working Group: How to Improve?" FNI Report 2. Available from https://www.fni.no/publications/challenges-of-the-arctic-council-s-sustainable-development-working-group-how-to-improve-article1417-290.html. Last visited 25 April 2019.

9. See https://oaarchive.arctic-council.org/bitstream/handle/11374/2092/%20SAOFI%20201%20_%202017%20_%20OULU_%20Info-%20Doc-%2001%20-%20B_%20Amarok-%20Maxi-%20Report.pdf?sequence=2&isAllowed=y%20for%20the%20most%20recently%20updated%20summary. Last visited 25 April 2019.
10. For more about PSI, see https://www.nefco.org/news-media/publications-reports/general/psi-project-flow-attachement-1-psi-guidelines. Last visited 25 April 2019.
11. However, there are examples of attempts that have been made. One is the US review of whether the US has implemented the recommendations of the ministerial meetings in the Arctic Council in the period 1998–2015. Not surprisingly the opinion is that so far this has been successful. The degree of implementation is still difficult to measure and an external review would probably have been preferable. See Arctic Council Ministerial Declaration Recommendations, 1998–2015. United States Implementation. June 2017. Available from https://oaarchive.arctic-council.org/handle/11374/2028. Last visited 25 April 2019. An example of external evaluation is WWF's Arctic Scorecard. This is an attempt to assess whether the Arctic States follow up on their commitments in the North. See http://wwf-ap.org/apps/acscorecard/. Last visited 25 April 2019.
12. Paula Kankaanpää and Oran Young (2012): "The effectiveness of the Arctic Council". We can also find other examples: Thomas Axworthy, Timo Koivurova and Waliul Hasanat (eds.) (2012) The Arctic Council: Its Place in the Future of Arctic Governance. Collections of papers. Munk-Gordon Arctic Security Program and the University of Lapland and Lassi Heinien, Heather Exner-Pirot and Joël Plouffe (2016) The Arctic Council: 20 Years of Regional Cooperation and Policy-Shaping, Arctic Yearbook.
13. Finland (2013), Finland's Arctic Strategy.
14. The Arctic parliamentary cooperation consists of parliamentarians from the eight Arctic countries and the European Parliament.
15. In the summer of 2018, Trump and Putin met in Finland, with no Arctic issues at the top of the agenda. The Finns still have ambitions to organise more regular summits around Arctic issues, but how this will be connected to the Arctic Council is unclear. Due to a cooler climate between Russia and the US, such formalised summits may seem somewhat remote, and when the US has a president with a strategy that can sometimes be difficult

to decipher, conditions are not quite right. If the Finns nevertheless succeed in arranging such a more formalised event, it is precisely because Finland being "neutral" is in a fairly unique situation. Furthermore, Arctic cooperation is an area with a tradition of low tension, and it is therefore natural to look to the most important Arctic cooperation, namely the Arctic Council.

16. Joint Memorandum of a Multilateral Audit on the Arctic states' National Authorities' Work with the Arctic Council. 2015.
17. CAFF, Actions for Arctic Biodiversity 2013–2021. Implementing the recommendations of the Arctic Biodiversity Assessment. Available from https://www.caff.is/actions-for-arctic-biodiversity-2013-2021. Last visited 25 April 2019.
18. Pekka Haavisto, Review of the Arctic Council Structures.
19. It should also be noted that it is a tradition that the leaders of the working groups meet in the capitals or other non-Arctic cities. Meetings related to various projects under the Arctic Council held virtually worldwide.
20. The two most well-known Arctic conferences, "Arctic Frontiers" and "Arctic Circle", will of course see this as very serious competition, but without dismissing the two mentioned examples, the Arctic Council has a completely different body of experience. It should be exploited better.
21. "Governance" is a frequently used term, although somewhat poorly defined. In general we can say that it applies to all the rules, norms, power relations and so on that govern a given field or social system, for example, the Law of the Sea, the states, the Arctic Council and so on will all be part of what we can call "Arctic governance".
22. An introductory work here (on the Norwegian national administration) is Svein Vigeland Rottem, Ida Folkestad Soltvedt and Christian Prip (2018) "Arctic Council in national Administration."

CHAPTER 5

The Arctic Council: Between Environmental Protection and Geopolitics

Abstract In the final and shortest chapter, I recapitulate the main arguments in this book and ask why we have an Arctic Council in the first place, emphasizing its role between environmental protection and geopolitics. By then, I hope the reader will have gained some understanding of the Arctic Council.

Keywords Arctic Council • Environmental protection • Geopolitics

In this book, the ambition has been to provide an overall synopsis of the history and work of the Arctic Council. Initially, the Council's path from limited environmental policy cooperation to the most important international cooperation in the Arctic was described. In parallel with this development of the Arctic Council, the Arctic has received increasing attention. One of the main reasons is the large climatic changes that are opening up new opportunities and challenges. Despite the fact that the Arctic petroleum hype has been toned down and we now rarely read analyses concluding that Arctic sea lanes will soon outstrip the Suez Canal and the Panama Canal, the Arctic is still on the political agenda both in the Arctic and in states far away from the Arctic Ocean. The most symbolic action was when a Russian research vessel planted a flag on the seabed at the North Pole.[1] This action was seen by many as confrontational and nurtured an emerging opinion that the Arctic would be the scene of a global power struggle for

S. V. Rottem, *The Arctic Council*,
https://doi.org/10.1007/978-981-13-9290-0_5

territory and resources. This was far from the truth, and this conflict discourse has lost some of its foothold. Still, it pops up regularly. The Arctic Council must deal with this narrative of where the region is heading. What the member states have done is to argue that the Council is an arena in which common challenges and opportunities will be discussed. The Council should not become an arena for political power struggles. It should act as an alternative place to meet in a geopolitically strained time. It has so far made good ground in this regard.

After drawing this broad picture in Chap. 1, the organisation of the Council and fundamental features of the attitudes of the various Arctic states to the cooperation were presented in Chaps. 2 and 3. Two key understandings became apparent. The agendas of the States are different, and there is not one "Arctic". As an extension of this we are seeing that the task portfolio of the Arctic Council is expanding. The question is whether there is also a limit for the Arctic Council. In continuation of this discussion, I ventured in Chap. 4 to make some recommendations for the road ahead. The underlying question is what is politically possible and makes practical sense. It is important to acknowledge that the Arctic Council is only one of the elements of Arctic governance. The states will be the main actors and the Law of the Sea is the key legal framework. This means that given that the Arctic Council is not undergoing a total transformation into a treaty-based international organisation, the Council's main role will be to contribute to knowledge and recommendations (and in some cases constitute a framework for negotiating internationally binding agreements) on developments in the Arctic, which can then be utilised at global, regional, national and local level. The Arctic Council is primarily a supplier of knowledge and not an international organisation with strong mechanisms for following up on binding decisions. The big and crucial question is therefore how to best use the knowledge about the Arctic that has been developed within the framework of the Arctic Council, so that through implementation and the transfer of knowledge, this knowledge has an impact on practical policy design. There is, however, more to the work of the Council.

* * *

All the Arctic states point out the Arctic Council as the most important cooperative body in the North, albeit often without a completely clear profile. Firstly, it can be argued that it is strategic realpolitik considerations that underlie this active approach to the Arctic Council. By supporting multilateral cooperation mechanisms, the Arctic states can constrain any expanding ambitions of non-Arctic states in the region. Demonstrating well-functioning cooperation mechanisms in the region also helps to restrain the conflict-oriented discourse we have seen regarding developments in the Arctic. The Arctic Council can also curb any competing regimes in the area, and ultimately the idea of an Arctic treaty modelled on the Antarctic Treaty. This was evident among other things in the debate on the incorporation of new observers. The argument is and was that by adopting a positive attitude towards the involvement of non-Arctic states in the region (through observer status in the Arctic Council), one has to a greater extent a firmer hand on the tiller. For example, the former Norwegian Minister of Foreign Affairs, Espen Barth Eide, stated in advance of the Kiruna meeting: "I often say that it is better that they want to join our club, than that they form a different club."[2]

At the same time and in continuation of the argument above, for the coastal States a particularly important point is that the Arctic Council supports the primacy of the Convention on the Law of the Sea in the Arctic Ocean, which gives them unique advantages. This is particularly visible by the fact that in order to become an observer (states, international organisations, NGOs) one must recognise the rights of the coastal states in the area.[3] Put bluntly and purely strategically, the cooperation itself and its framework are more important than the actual content of the cooperation. The strategy is in the form not the content. Through active support for the Arctic Council, the Arctic states deny any competing forms of governance for the region, for example through the UN. The Arctic states want control over developments in the region (management, resource utilisation, commercial activity, etc.). By referring to the work of the Council they say that the Arctic is not an unregulated no-man's land.

At the same time, this foreign policy commitment can be seen as habit.[4] The Arctic states join together wherever possible, so long as it is not too binding and there is talk of international cooperation. This does not have to be based on specific strategic considerations, but rather a fundamental positive attitude towards international cooperation in itself. Furthermore, there are major national differences, and the member states will have different answers to the question: Why an Arctic Council? For the US, it is important to be involved wherever the Russians are while at the same time the argument that the Council has a stabilising function is important. In the case of Russia, it is important to appear as a legitimate and important actor in the North. The Arctic Council is a good tool in this respect. Of course, there are also goals that are more limited and idealistic goals might also important. The cooperation facilitates a better understanding of the climate and environmental changes in the North and makes it possible to make wise choices in the Arctic. Such more limited goals are also found in the smaller states. For Canada, the Arctic Council has been an important platform for including indigenous peoples, and in Norway, the Arctic Council has been important for strengthening the northern part of the country. This is apparent through the active work in getting the permanent Secretariat established in Tromsø and bringing the secretariats of the ACAP, AMAP and EPPR working groups together. For Norway it has also been important to have good relations with their bigger neighbour in the north.[5] Iceland, Denmark, Sweden and Finland have at times been less active in the Arctic Council, but here too there is a desire to put a national mark on what is important in the North, for example the preservation of the Arctic environment, but also to demonstrate relevant expertise for supporting sustainable development.

The Arctic Council is also an important arena for networking that has allowed many people to be heard. Despite challenges in terms of, for example, funding, indigenous groups are regarded as legitimate actors. Non-Arctic States can contribute their competence in the working groups. NGOs such as WWF are included and can put pressure on political decisions. And last but not least, researchers and employees of the national administrations have been given a meeting place for increasing their knowledge of what is happening in the Arctic. These are among the main reasons why we should continue to have an Arctic Council. And, while in all the capitals of the Arctic states the Arctic Council is being described as the most important forum for cooperation, and the list of states with ambitions in the regions is getting longer and longer, it is a good idea to have an insight into what this forum really is and can be. This is what I hope this book has helped with.

Notes

1. At the same time, Russian Foreign Minister Sergei Lavrov emphasised that this was a research expedition and not an attempt to annex the pole: "When explorers reach a point where no one has been before, they plant a flag." Cited in Charles Emmorson (2010): *The Future History of the Arctic* (New York: Public Affairs), p. 82.
2. Aftenposten (2013) "Norway will give China a place in the Arctic Council". 12 May 2013.

3. Arctic Council 2013, The Arctic Council Observer Manual (last updated 2016).
4. See, for example, Ted Hopf (2010): "The logic of habit in International relations". *European Journal of International Relations*, 16 (4), pp. 539–561.
5. In that regard one can cite researcher and Arctic Council expert Elana Rowe: "Norway's status as an information provider, a convener and a bridge to Russia gives the country a degree of influence in Arctic multi-lateral settings" Elana Rowe (2013): "Arctic hierarchies? Norway, status and the high north". *Polar Record*. Published online February 2013, p. 1.

Index

S. V. Rottem, *The Arctic Council*,
https://doi.org/10.1007/978-981-13-9290-0

www.ingramcontent.com/pod-product-compliance
Ingram Content Group UK Ltd.
Pitfield, Milton Keynes, MK11 3LW, UK
UKHW021843060925
462651UK00010B/109

* 9 7 8 9 8 1 1 3 9 2 9 2 4 *